WITHDRAWN

C0-DKN-188

DATE DUE

Demco, Inc. 38-293

MAR 2 2 2012

WITHDRAWN

DISCARD

MEDIA AND COMMUNICATIONS – TECHNOLOIGES, POLICIES
AND CHALLENGES

MEDIA INDUSTRY PROGRAMMING, COMPETITION AND COPYRIGHT ISSUES

MEDIA AND COMMUNICATIONS –
TECHNOLOIGES, POLICIES AND CHALLENGES

Additional books in this series can be found on Nova's website
under the Series tab.

Additional E-books in this series can be found on Nova's website
under the E-books tab.

Columbia College Library
600 South Michigan
Chicago, IL 60605

MEDIA AND COMMUNICATIONS – TECHNOLOIGES, POLICIES
AND CHALLENGES

MEDIA INDUSTRY PROGRAMMING, COMPETITION AND COPYRIGHT ISSUES

RYAN E. MOORE
EDITOR

Nova Science Publishers, Inc.
New York

Copyright ©2011 by Nova Science Publishers, Inc.

All rights reserved. No part of this book may be reproduced, stored in a retrieval system or transmitted in any form or by any means: electronic, electrostatic, magnetic, tape, mechanical photocopying, recording or otherwise without the written permission of the Publisher.

For permission to use material from this book please contact us:
Telephone 631-231-7269; Fax 631-231-8175
Web Site: http://www.novapublishers.com

NOTICE TO THE READER

The Publisher has taken reasonable care in the preparation of this book, but makes no expressed or implied warranty of any kind and assumes no responsibility for any errors or omissions. No liability is assumed for incidental or consequential damages in connection with or arising out of information contained in this book. The Publisher shall not be liable for any special, consequential, or exemplary damages resulting, in whole or in part, from the readers' use of, or reliance upon, this material. Any parts of this book based on government reports are so indicated and copyright is claimed for those parts to the extent applicable to compilations of such works.

Independent verification should be sought for any data, advice or recommendations contained in this book. In addition, no responsibility is assumed by the publisher for any injury and/or damage to persons or property arising from any methods, products, instructions, ideas or otherwise contained in this publication.

This publication is designed to provide accurate and authoritative information with regard to the subject matter covered herein. It is sold with the clear understanding that the Publisher is not engaged in rendering legal or any other professional services. If legal or any other expert assistance is required, the services of a competent person should be sought. FROM A DECLARATION OF PARTICIPANTS JOINTLY ADOPTED BY A COMMITTEE OF THE AMERICAN BAR ASSOCIATION AND A COMMITTEE OF PUBLISHERS.

Additional color graphics may be available in the e-book version of this book.

LIBRARY OF CONGRESS CATALOGING-IN-PUBLICATION DATA

Media industry programming, competition and copyright issues / editor, Ryan E. Moore.
     p. cm.
   Includes index.
   ISBN 978-1-61122-078-0 (hbk.)
   1. Broadcasting--United States. 2. Television programs--United States. 3. Radio programs--United States. 4. Broadcasting--Ownership--United States. I. Moore, Ryan E.
   HE8689.8.M43 2010
   384.540973--dc22
                          2010041305

Published by Nova Science Publishers, Inc. ✚ *New York*

CONTENTS

PREFACE

The media industry plays a vital role in informing and entertaining the public. Media ownership and the availability of diverse programming have been a long-standing concern of Congress. Despite numerous programming choices in television and radio available to the public, some studies have reported that independently produced programming, that is, programming not affiliated with broadcast networks or cable operators, has decreased through the years. This book discusses the extent to which the sources of television programming have changed over the last decade; the factors industry stakeholders identified as affecting the availability of independent television programming; and the factors industry stakeholders identified as influencing decisions in radio.

Chapter 1- The media industry plays a vital role in informing and entertaining the public. Media ownership and the availability of diverse programming have been a longstanding concern of Congress. Despite numerous programming choices in television and radio available to the public, some studies have reported that independently produced programming—that is, programming not affiliated with broadcast networks or cable operators—has decreased through the years. This requested report discusses (1) the extent to which the sources of television programming have changed over the last decade, (2) the factors industry stakeholders identified as affecting the availability of independent television programming, and (3) the factors industry stakeholders identified as influencing programming decisions in radio. To address these issues, GAO analyzed data from the Federal Communications Commission (FCC) and industry on sources of broadcast television programming in prime time (weeknights generally from 8:00 p.m. to 11:00 p.m.) and companies owning cable networks, as well as radio format data to determine programming variety. GAO also reviewed legal, agency, and industry documents and interviewed industry stakeholders, public interest groups, and others.

GAO provided FCC with a draft of this report for comment. In response, FCC provided technical comments that the authors incorporated where appropriate.

Chapter 2- The Satellite Television Extension and Localism Act of 2010 (STELA), P.L. 111-175, modifies the copyright and carriage rules for satellite and cable retransmission of broadcast television signals. The legislation was needed to reauthorize (through December 31, 2014) certain expiring provisions in the Copyright Act and the Communications Act and to update the language in those acts to reflect the transition from analog to digital transmission of broadcast signals, as well as to address certain public policy issues. Had the expiring provisions not been reauthorized, satellite operators would have lost access to a statutory compulsory copyright license and to statutory relief from retransmission consent requirements. This would have made it difficult, if not impossible, for them to retransmit certain distant broadcast signals to their subscribers, including signals providing otherwise unavailable broadcast network programming.

Chapter 3- In December 2007, the Federal Communications Commission relaxed its newspaper/broadcast ownership ban (order released February 2008). The decision raised concerns in Congress about increasing media consolidation that have long been at the forefront of the debate over ownership restrictions. The Commission's order served to rekindle the discussion of media consolidation and the perceived need to take action to preserve a diversity of voices in the marketplace of ideas. The FCC rule, as this report illustrates, has a history dating back to a previous failed attempt to relax a greater number of broadcast cross-ownership restrictions, and it is worthwhile to examine this previous proceeding in order to understand the current status of the rules.

Chapter 4 – This is a statement of Brian L. Roberts, Chairman and Chief Executive Officer, Comcast Corporation and Jeff Zucker, President and Chief Executive Officer, NBC Universal, before the Committee on the Judiciary, United States House of Representatives.

Chapter 5 – This is a statement of W. Hazlett, Panel on the Comcast-NBCU Venture, before the United States House of Representatives, Judiciary Committtee Hearings.

Chapter 6 - This is a statement of Mark Cooper, Director of Research, Consumer Federation of Amercia on behalf of Consumer Federation of

America free Press Consumers Union, before the United States House of Representatives, Committee on the Judiciary, Subcommittee on Antitrust, Competition Policy and Consumer Rights.

Chapter 7 - This is a statement of Larry Cohen, President, Communications Workers of America, before the United States House of Representatives, Committee on the Judiciary.

Chapter 8 - This is a statement of Andrew Jay Schwartzman, president and CEO, Media Access Project, before the Committee on the Judciary, United States House of Represenatives.

In: Media Industry Programming, Competition...   ISBN: 978-1-61122-078-0
Editors: Ryan E. Moore   ©2011 Nova Science Publishers, Inc.

Chapter 1

MEDIA PROGRAMMING: FACTORS INFLUENCING THE AVAILABILITY OF INDEPENDENT PROGRAMMING IN TELEVISION AND PROGRAMMING DECISIONS IN RADIO

United States Government Accountability Office

WHY GAO DID THIS STUDY

The media industry plays a vital role in informing and entertaining the public. Media ownership and the availability of diverse programming have been a longstanding concern of Congress. Despite numerous programming choices in television and radio available to the public, some studies have reported that independently produced programming—that is, programming not affiliated with broadcast networks or cable operators—has decreased through the years. This requested report discusses (1) the extent to which the sources of television programming have changed over the last decade, (2) the factors industry stakeholders identified as affecting the availability of independent television programming, and (3) the factors industry stakeholders identified as influencing programming decisions in radio. To address these issues, GAO analyzed data from the Federal Communications Commission (FCC) and industry on sources of broadcast television programming in prime time (weeknights generally from 8:00 p.m. to 11:00 p.m.) and companies owning cable networks, as well as radio format data to determine programming variety. GAO also reviewed legal, agency, and industry documents and interviewed industry stakeholders, public interest groups, and others.

GAO provided FCC with a draft of this report for comment. In response, FCC provided technical comments that we incorporated where appropriate.

WHAT GAO FOUND

The sources of broadcast and basic cable television programming have changed little in recent years. As a source of programming for prime time television, major broadcasters (ABC, CBS, Fox, and NBC) and their affiliated studios produced the majority of programming in each of the selected years that GAO analyzed. In particular, GAO found major broadcasters produced about 76 to 84 percent of prime time programming hours. The remaining programming came from independent producers, which are not affiliated with the major broadcasters. Since basic cable networks are also a source of television programming, GAO analyzed the ownership of those networks as an indicator of which entities control the television programming. On the basis of GAO analysis of ownership in the 20 most widely distributed basic cable networks, major broadcasters and companies affiliated with both major broadcasters and cable operators have owned half or more of the top 20 cable networks for each year reviewed. Combining ownership in both prime time programming and basic cable networks, the major broadcasters have controlled a significant share of television programming over the last decade.

Stakeholders primarily cited economic factors as influencing the availability of independent television programming. In this regard, producers GAO contacted stated that developing and producing broadcast television programs is costly and financially risky. And while funds need to be secured early on in the development and production process to finance these costs, independent producers stressed that it is difficult to obtain financing for production costs. For cable television (viewed through a subscription video service), representatives of independent cable networks said a new network faces considerable uncertainty as to whether it will be distributed by a sufficient number of video providers (such as Comcast and DirecTV) to make its operations viable. By contrast, cable networks developed by cable operators or major broadcasters are able to negotiate distribution of the network with video providers as part of an agreement for distribution of an established affiliated network.

For radio, stakeholders cited economic factors, local community interests, and consolidation in the radio industry as influences on programming decisions. Among both commercial and public radio stations, stakeholders said that programming decisions are based on listeners' interests in a given market. GAO found that within two of the three largest local markets nationwide, many of the most common local radio formats differ from the most common radio formats nationally, indicating that programming decisions are affected by local community interests. Over the last 10 years there has been consolidation in the radio industry; however, stakeholders' opinions varied about the extent to which consolidation has affected programming decisions. While some studies show that consolidation has led to homogenized radio playlists in different markets nationwide, GAO's analysis shows diverse formats and preferences are reflected within individual local markets.

ABBREVIATIONS

BIA*fn*	Broadcast Investment Analyst Financial Network
FCC	Federal Communications Commission
MSA	Metropolitan Statistical Area
NPR	National Public Radio

March 17, 2010

The Honorable Patrick J. Leahy
Chairman
Committee on the Judiciary
United States Senate

The Honorable Herbert Kohl
Chairman
Subcommittee on Antitrust, Competition Policy, and Consumer Rights
Committee on the Judiciary
United States Senate

The Honorable Byron L. Dorgan
United States Senate

The media industry plays an important role in educating and entertaining the public. Given this vital role, the ownership of media outlets and the availability of diverse programming in the media have been a long-standing concern of Congress. One such concern, in particular, is that with consolidation in the media industry, the percentage of independently produced programming content on media outlets has decreased, thus limiting the number of distinct media voices and selection choices for the public.[1] In 1995, the Federal Communications Commission (FCC) repealed the Financial-Syndication Rules, thereby allowing broadcast networks to have ownership interests in television programs during prime time.[2] In addition, the Telecommunications Act of 1996, among other things, directed FCC to conduct a rulemaking to evaluate local limits of ownership in television and relaxed the ownership limits in local radio stations, leading to consolidation in some segments of the media industry.[3] For example, prior to the act, the largest radio station owner owned fewer than 65 radio stations, whereas in 2009, the largest radio station owner owned over 800 stations, raising concerns about the variety of programming available on the radio. Although the current media environment provides the public with numerous programming choices in television and radio, and over the Internet, some media industry stakeholders and studies have reported that independently produced programming has decreased through the years as a result of media consolidation.

You requested that we study the state of programming for television and radio and the factors influencing programming decisions. As such, we reviewed (1) the extent to which the sources of programming in television have changed over the last decade, (2) the factors and conditions industry stakeholders identified as affecting the availability of independent programming in television, and (3) the factors industry stakeholders identified as influencing programming decisions in radio.

To determine the extent to which the sources of programming in television have changed during the last decade, we analyzed available data on two key sources of television programming—companies producing prime time television programs for major broadcast networks[4] and companies with basic cable network[5] ownership interests over the last decade. We focused on programs broadcasted on prime time because that is the block of time on television with generally the most viewers, and in turn these programs generate the most advertisement revenue for networks.[6] To determine which companies produced prime time broadcast television programming, we classified prime time programs into two categories: (1) programs produced by major broadcasters, and (2) programs produced by independent production companies not affiliated with a major broadcaster (independent producers). Because annual data that track program production in these categories are limited, we analyzed the fall prime time schedules in 2002, 2005, 2008, and 2009 and classified them in the two categories. We selected these years based on available data from FCC's previous study that contained data in the two categories for 2002. We then conducted our analysis for every third year and classified the programs into the two categories. We also analyzed the 2009 fall prime time schedule to provide the most current data available. Since basic cable networks are also a source of television programming, we analyzed the ownership of those networks as an indicator of which entities control the television programming on the networks. Specifically, we used data from SNL Kagan to determine the types and the number of companies with ownership interests in basic cable networks that were available from 1998 to 2008, and the companies that owned the largest number of basic cable networks during this period.[7] Of the 20 most widely distributed basic cable networks, as measured by the number of subscribers for each year from 1998 to 2008, we determined how many were affiliated with major broadcasters, cable operators, and other media companies and how many were unaffiliated independent cable networks.

To determine the factors industry stakeholders identified as affecting the availability of independent programming in television, and as influencing programming decisions in radio, we met with FCC officials and interviewed or obtained written comments from selected academic experts and industry stakeholders and associations, including representatives from broadcasting (such as ABC and Fox), cable and satellite (such as Comcast and DirecTV), commercial and public radio (such as Clear Channel Communications and National Public Radio), independent programming groups (such as Future of Music Coalition and Independent Film and Television Alliance), and public interest groups (such as Consumers Union and Free Press). See appendix I for a complete list of academic experts and industry stakeholders we contacted. We selected experts and industry stakeholders based on published studies, representation of the different segments of the media (i.e., broadcast and cable televisions and radio), and recommendations from other industry stakeholders; we intended to obtain diverse views and did not weight views of the experts and stakeholders but grouped similar stakeholders that represent a segment of the media industry. We also reviewed the relevant laws, regulations, and literature, including comments filed by stakeholders in various FCC proceedings. Additionally, for radio, we focused on radio station formats indicating the types of programming a station might play.[8] To conduct this analysis, we obtained historical data on the distribution of radio stations by their primary formats nationwide and in local markets from 1999 to 2003 and station-level format data from the Broadcast Investment Analyst Financial Network's (BIA*fn*) Media Access Pro Database for commercial and public radio stations from 2004 to 2009. Although the BIA*fn* format data provide a general overview of the type of programming aired on a given radio station, they do not identify specific programming content that is played on the station. We did not assess independently produced programming on radio because a national playlist database identifying record label affiliation is not available. We analyzed the format data to determine programming variety and distribution of radio stations by their format nationwide and in two selected local markets in 2009.[9] For local markets, we selected New York and Chicago because they are similar in size with different demographic populations. We also examined 2009 format data for commercial and public radio stations and identified the top 10 most popular formats (based on the number of stations with the particular formats available) for each group nationwide.[10] To examine the extent of programming variety among radio stations, we selected the top 10 radio station owners—that is, owners who own the most radio stations nationwide—and examined similarities and differences in formats for each owner's radio stations in the same market. To identify the top 10 radio station owners in 1996-1998, 2000-2002, 2007, and 2009 we used data from FCC reports and the BIA*fn* database. The top 10 radio station ownership data were not available in 2003-2006 and 2008. We also reviewed studies on radio programming for information on radio station playlists and the extent to which playlists for each owner's radio stations overlap in the same market. We determined that the television and radio data we obtained were sufficiently reliable for the purposes of this report.

We conducted our work from May 2009 to March 2010 in accordance with all sections of GAO's Quality Assurance Framework that are relevant to our objectives. The framework requires that we plan and perform the engagement to obtain sufficient and appropriate evidence to meet our stated objectives and discuss any limitations in our work. We believe that the information and data obtained, and the analysis conducted, provide a reasonable basis for any findings and conclusions in this product. Appendix I contains a more detailed discussion of our objectives, scope, and methodology.

Table 1. Glossary of Terms

Term	Definition
Basic cable network	An organization that may produce television programs, which are distributed to the public through a subscription video service.
Independent cable network	An organization that is not affiliated with major broadcasters, cable operators, or satellite providers.
Radio station format	Type of programming content on a radio station, such as Adult Contemporary, Country, Jazz, News and Information, Sports, Talk, and so forth.
Public radio stations	Locally owned and operated stations that receive some or all of their funding from listener contribute-ons, the federal government, or other sources. Some public radio stations are affiliated with National Public Radio, which is a national radio service that provides station programming content.
Multichannel video programming distributors (referred to as video providers in this report)	Entities such as a cable operator, a direct broadcast satellite service, or telecommunications company that distributes video programming to subscribers for a subscription fee.
Scripted programming	Programs that are developed based on written scri-pts by writers and producers and include different program genres, including comedy and drama.
Nonscripted programming	Programs that are not based on written scripts, such as reality programs, game shows, and sports.
Arbitron radio market	A geographically contiguous area in which the listenership of radio stations is surveyed for ratings by Arbitron Inc.

Source: GAO based on FCC and industry information.

BACKGROUND

The media industry has its own terminology, and the following glossary provides the definition of terms used throughout this report:

Typically, the general public views television programming through broadcast or subscription video service. Broadcast television provides free over-the-air programming to the public through local television stations. By contrast, consumers pay fees for subscription video service to video providers, including cable operators, satellite providers, or telecommunications companies. Programming for broadcast and subscription video service differs, as illustrated in figure 1.

Broadcast television consists mainly of four major broadcast networks (ABC, CBS, Fox, and NBC) and several smaller networks, such as the CW Television Network, MyNetworkTV, and ION Television. Each of the four major broadcasters owns and operates some local television stations; other stations can be affiliated with one of the major broadcasters or, as is the case with public television, unaffiliated with the major broadcasters.[11] The four major broadcasters provide scripted and nonscripted programming to the local television stations that is produced either by the major broadcasters' affiliated production companies or by independent producers. The development process of scripted programs (i.e., drama and comedy series) for prime time programming involves steps that allow major broadcasters to periodically assess the program as it develops, as described in figure 2.

In contrast, the development process for nonscripted programs, such as reality programs and game shows, does not involve most of the steps shown in figure 2. Scripts and pilots do not need to be developed for nonscripted programs, making them less expensive to produce than scripted programs.

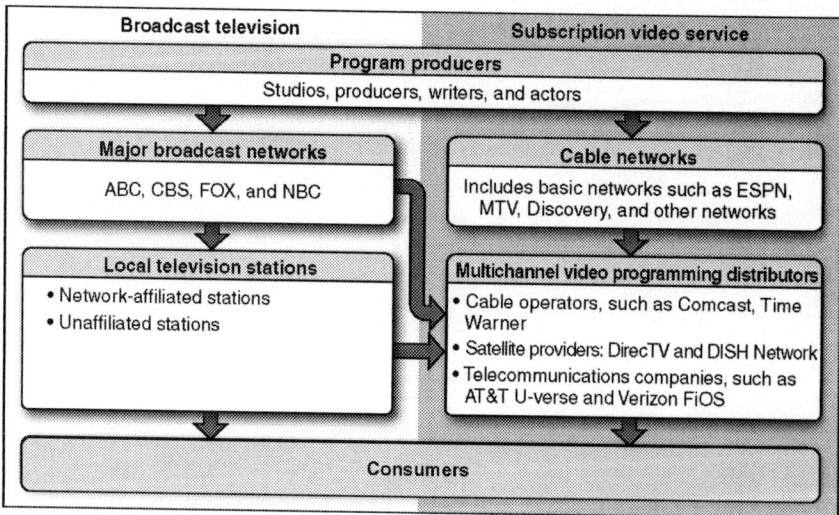

Source: FCC and GAO.

Figure 1. Television Programming in Broadcast and Subscription Video Service

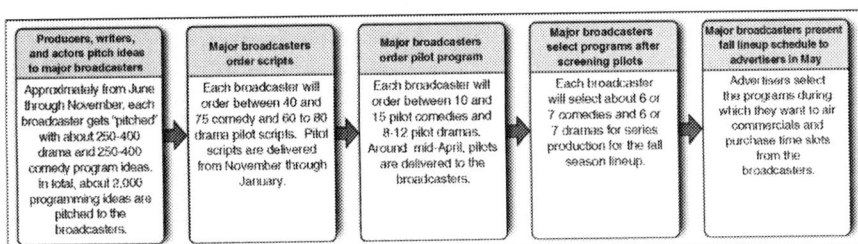

Producers, writers, and actors pitch ideas to major broadcasters	Major broadcasters order scripts	Major broadcasters order pilot program	Major broadcasters select programs after screening pilots	Major broadcasters present fall lineup schedule to advertisers in May
Approximately from June through November, each broadcaster gets "pitched" with about 250-400 drama and 250-400 comedy program ideas. In total, about 2,000 programming ideas are pitched to the broadcasters.	Each broadcaster will order between 40 and 75 comedy and 60 to 80 drama pilot scripts. Pilot scripts are delivered from November through January.	Each broadcaster will order between 10 and 15 pilot comedies and 8-12 pilot dramas. Around mid-April, pilots are delivered to the broadcasters.	Each broadcaster will select about 6 or 7 comedies and 6 or 7 dramas for series production for the fall season lineup.	Advertisers select the programs during which they want to air commercials and purchase time slots from the broadcasters.

Source: GAO analysis of major broadcasters' data.

Figure 2. Annual Broadcasting Program Development Process for Scripted Programming

Table 2. Top 10 Radio Station Owners in 2009 and Number of Stations Owned in 2007-2009

Radio station owners	2007	2008	2009
Clear Channel Communications Inc.	1,135	694	847
Cumulus Broadcasting LLC	306	306	305
Educational Media Foundation	180	205	253
Citadel Communications	212	204	205
American Family Association Inc.	126	130	137
CBS Radio	140	140	134
Entercom	120	115	112
Salem Communications Corporation	98	97	93
Saga Communications Inc.	89	91	91
Cox Radio Inc.	79	79	85

Source: GAO analysis of BIA*fn* data.

For subscription video service, video providers obtain a variety of programming from both broadcasters (which can include major networks and local stations) and cable networks.[12] Video providers must negotiate with broadcasters and cable networks to air and distribute their programming. Negotiations include the price, terms, and conditions for distribution on the video providers' systems. Video providers have the discretion to select which cable networks will be available and, subject to negotiation, how they will be packaged and marketed to subscribers.[13]

According to a recent FCC report, more than 500 cable networks exist, including national cable networks (such as CNN, Discovery Channel, ESPN, and Fox News) as well as regional cable networks (such as the California

Channel, Comcast SportsNet Chicago, and the YES Network). Cable networks can provide niche programming—that is, programming that targets specific demographics. For instance, Lifetime Network offers programming that specifically targets women, while MTV Network targets programming for the 18-to-34 age demographic.

The general public receives radio programming through commercial and public radio stations. Over the last 5 years, the number of full-power radio stations has increased from 13,590 in 2005 to over 14,600 in 2009, with the vast majority of these stations being commercial (78 percent, or 11,430 stations) and the remainder being public (22 percent, or 3,198 stations).[14] Following passage of the Telecommunications Act of 1996, concentration in radio station ownership increased significantly because of the act's relaxation of national and local multiple radio ownership limits.[15] For example, in 1996, the two largest radio station owners held fewer than 65 radio stations each. By contrast, as of 2009, Clear Channel Communications Inc. owned over 800 radio stations (down from 1,135 in 2007), and the second largest group owner, Cumulus Broadcasting LLC, owned about 300 radio stations (see table 2).[16] In 2009, the top 10 radio station owners owned 20 percent of all commercial radio stations. In addition, each radio station has a primary programming format designation that describes the programming content on that station. For example, in 2009, radio station KQSD in Lowry, South Dakota's, primary format was Classical, its secondary format was News, and its tertiary format was Jazz. As such, the station primarily plays Classical music, but it also provides some news and plays some Jazz.

FCC awards licenses to television and radio stations to use the airwaves expressly on the condition that licenses serve the public interest and licensees are responsive to the needs of its local community. Toward this end, FCC has long identified localism, competition, and diversity as its three core goals of media policy. Within this framework, FCC has considered the public interest best served by promoting free expression of diverse views and has promoted program diversity by limiting the number of broadcast outlets any one entity may own. As such, individual radio and television stations generally have discretion to select programming and to determine how best to serve the local community audience.

Since the mid-1990s, FCC has amended or repealed a number of rules and regulations affecting the media industry. In 1995, FCC repealed the Financial Interest and Syndication Rules (Fin-Syn rules)[17] so that a major broadcaster can own programming that it airs during prime time hours, as well as own syndication rights to programs purchased from independent producers.[18]

Following the repeal of the Fin-Syn rules, each of the four major broadcasters merged with, or acquired an ownership interest in, at least one major production studio. For instance, the Walt Disney Company acquired ABC and developed ABC Television Studio; CBS became affiliated with the studio Paramount Television; and NBC merged with Universal Pictures. In addition, News Corporation—which launched the Fox Broadcasting Network in 1986—owns several production studios, including 20th Century Fox. FCC is required to review media ownership rules every 4 years and determine whether those rules are necessary in the public interest.[19]

Although FCC regulates television primarily through ownership rules and station licensing, some of its other rules also affect aspects of television programming. Some of the key rules that affect programming and carriage were adopted in 1992 and are summarized below.[20]

- **Retransmission consent and must carry rules.**[21] Under these rules, every 3 years local commercial television stations (including those owned and operated by the major broadcasters) must decide whether to negotiate individual retransmission consent agreements with each cable operator in its designated market area for compensation in exchange for the cable operator's right to carry the broadcast signal.[22] In lieu of negotiation, stations may elect to require each cable operator in its designated market area to carry its signal (i.e., must carry), without receiving compensation for such carriage.
- **Program carriage rule.**[23] This rule prevents a video provider from requiring a financial interest in programming or coercing a programmer (i.e., cable network) to grant exclusive rights as a condition for carriage, or from discriminating against an independent cable network in a way that unreasonably restrains the ability of the network to compete.
- **Commercial leased access rule.**[24] Under this rule, cable operators are required to set aside a certain number of channels, depending on the size of the cable system, that can be leased out to independent cable networks for access on its distribution system.[25] Congress has required FCC to (1) determine the maximum reasonable rates that a cable operator may establish for commercial use of the designated channels; (2) establish reasonable terms and conditions for such use, including those for billing and collections; and (3) establish procedures for the expedited resolution of disputes concerning rates or carriage.[26]

Percentage of broadcast prime time program hours

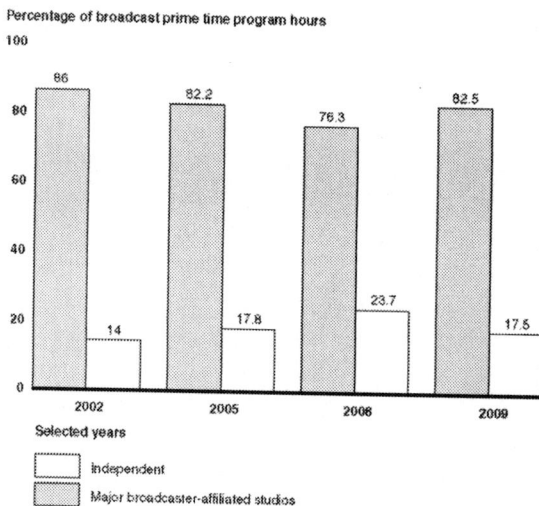

Source: GAO analysis of data from *FCC Media Ownership Working Group Report* and *International Television & Video Almanac* used by permission from Quigley Publishing Company.

Note: For each of the selected years, the independent category includes studios, regardless of their size, that were not affiliated with a major broadcaster. For example, Universal Television Studios merged with NBC in 2004; therefore it was included in the independent category in 2002 and in the broadcaster category in the other selected years. Sony Pictures Television Studio is not affiliated with a major broadcaster, so it is included in the independent category for shows it produced in the selected years.

Figure 3. Percentage of Broadcast Prime Time Program Hours Provided by Major Broadcaster-Affiliated Studios and Independent Producers for 2002, 2005, 2008, and 2009

SOURCES OF BROADCAST AND CABLE TELEVISION PROGRAMMING HAVE CHANGED LITTLE IN THE LAST DECADE

Major Broadcasters and Their Affiliated Studios Have Produced the Majority of Broadcast Prime Time Programming

Major broadcasters and their affiliated studios have produced the majority of broadcast prime time programming in each of the selected years that we analyzed. In particular, major broadcaster-affiliated studios produced from 76 to 84 percent of broadcast prime time programming hours, with the remaining hours coming from independent producers. As shown in figure 3, in most of the years that we reviewed, the share of major broadcaster-produced prime time programs did not change significantly. However in 2008, the prime time programming from independent producers increased slightly compared with such programming in 2005.

For the fall 2009 broadcast prime time schedule, the top five program producers as measured in prime time program hours were studios affiliated with ABC, CBS, Fox, NBC, and Warner Bros.[27] These producers provided approximately 76 total programs, amounting to about 82 percent, in the fall prime time schedule. We identified 11 prime time programs that fell into the independent producer category for the fall 2009 prime time schedule. Of those, Sony Pictures Television Studio produced 3 programs, and eight other independent producers each supplied a program. Although most of the programs produced during the years we reviewed were affiliated with major broadcasters, a previous FCC-commissioned study indicated that the number and affiliation of prime time programming producers has changed significantly since the repeal of the Fin-Syn rules in 1995.[28] The study found that in 1995, the top five program producers provided about 54 percent of prime time programming, with three producers affiliated with a major broadcaster.

Numerous Companies Own Cable Networks, but Major Broadcasters and Their Affiliated Companies Have Continued to Own about Half of the Most Widely Distributed Cable Networks

Since basic cable networks are also a source of television programming, we analyzed the ownership of those networks as an indicator of which entities control the television programming on the networks. On the basis of our analysis of ownership interests over the last decade, we found that a number of companies have ownership interests in a basic cable network (cable network), but a much smaller group of companies have ownership interests in 5 or more such networks.[29] From 1998 to 2008, 94 companies on average have owned an interest in at least 1 cable network. The number of companies has declined somewhat over time, however, from a high of 106 companies in 1998 to a low of 81 companies in 2008.[30] Cable network owners include owners of major broadcasters, such as News Corporation, which owns Fox, and Walt Disney

Company, which owns ABC; cable operators, such as Comcast and Cablevision; owners of major publications and television stations, such as Tribune Company and Hearst Corporation; and other media companies, such as Liberty Media Corporation and Scripps Networks Interactive. On the basis of our analysis of all the companies with cable network ownership interests from 1998 to 2008, we found a range of 11 to 13 companies that owned an interest in 5 or more networks in at least 1 year. Of these companies, we found a range of 5 to 7 companies that owned at least 12 cable networks over the decade. As shown in figure 4, the number of basic cable networks owned by these top 5 companies has not changed significantly over the last 11 years.

Number of cable networks

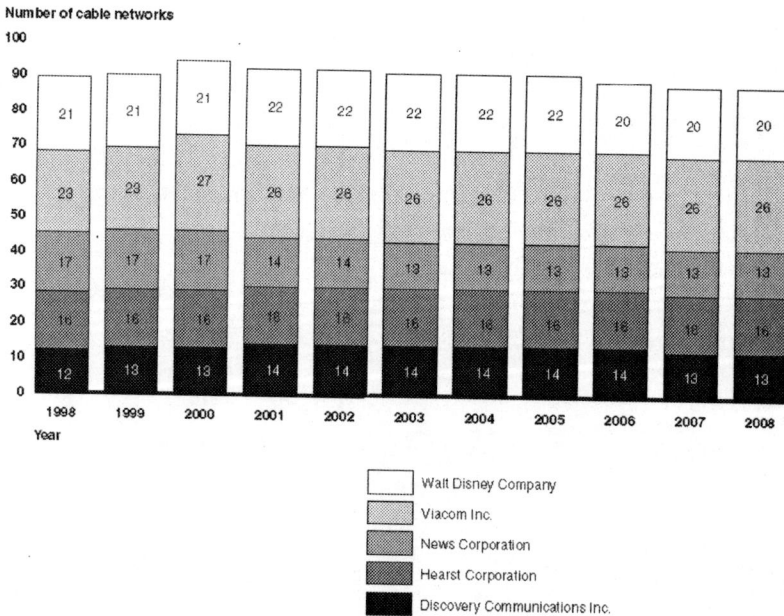

Source: GAO analysis of SNL Kagan data.

Note: The top owners were determined using company names presented in data obtained from SNL Kagan. For example, for 2008, the data indicate that NBC Universal had an ownership interest in 7 basic cable networks and General Electric (GE) had an ownership interest in 11 cable networks. Although GE is a parent company of NBC Universal, this analysis does not attribute ownership interest in the 7 NBC Universal-owned cable networks to GE. In 2008, the total number of cable networks with either a GE or NBC Universal affiliation totaled 18.

Figure 4. Top Five Owners of Basic Cable Networks and the Number of Cable Networks They Have Owned from 1998 to 2008

Number of cable networks

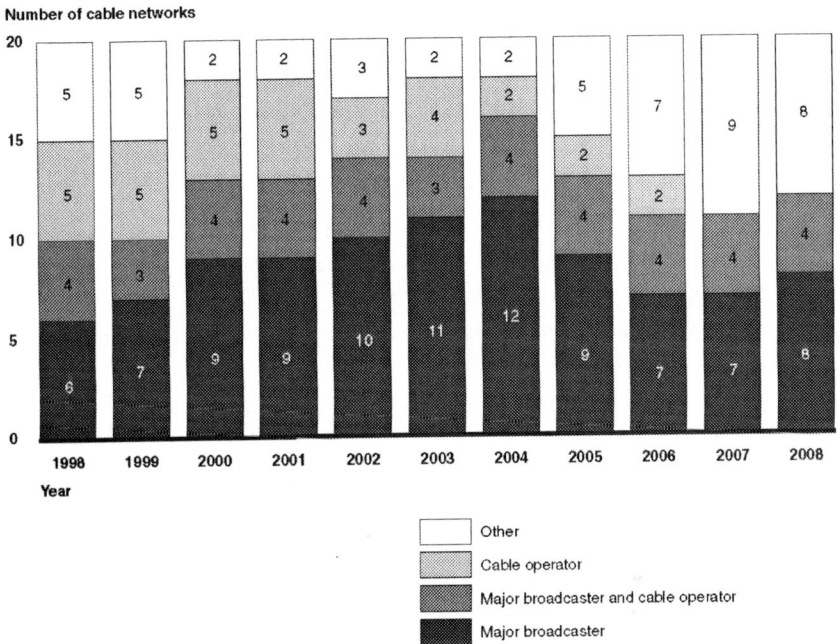

Source: GAO analysis of SNL Kagan data.

Note: We categorized owners that had an ownership interest in a top 20 basic cable network into four categories: major broadcaster, both major broadcaster and cable operator (or satellite provider), cable operator (or satellite provider), and other. Basic cable network owners that fell into the "other" category were generally those that could not be identified with the three other categories and included independent cable networks.

Figure 5. Number of Networks by Ownership Interest in the Top 20 Most Widely Distributed Basic Cable Networks

Viacom and Walt Disney Company had ownership interests in the most cable networks over the last decade, with each owning more than 20 networks in each year. None of the top five owners has increased the number of cable networks owned since 2001. In 2008, these top five companies owned about half of basic cable networks.

We analyzed ownership of the 20 most widely distributed basic cable networks, as measured by the number of subscribers for each year from 1998 to 2008 (referred to as top 20 cable networks). On the basis of our analysis, we found major broadcasters and companies affiliated with both major broadcasters and cable operators combined owned 50 percent or more of the top 20 networks. As shown in figure 5, the number of major broadcaster-owned top 20 cable networks ranged from 6 in 1998 to a high of 12 in 2004 before declining to 8 cable networks in 2008. The number of top 20 cable networks owned by companies affiliated with both major broadcasters and cable operators remained relatively steady during the decade at 3 to 4. Cable operators without a broadcast company affiliation owned 5 of the top 20 cable networks in 1998, but this number declined over time and was zero in 2007 and 2008.

In 2008, the last year of our analysis of ownership of the top 20 cable networks, we found 8 cable networks that were affiliated with major broadcasters. For example, 2 top 20 cable networks, ABC Family Channel and Disney Channel, are owned by Disney, a company that also owns the ABC broadcast network. Four networks in the top 20 were affiliated with both major broadcasters and cable operators. For example, in 2008, CNN, TBS, and TNT were owned by Time Warner, a company affiliated with cable operator Time Warner Cable, broadcaster CW Television Network, and television production studio Warner Bros.[31] In addition, 8 networks in the top 20 fell in the "other" category for 2008, because they did not appear to have a direct affiliation with a major broadcaster, cable operator, or satellite provider. Some of the networks in this category, including the Food Network and HGTV network, which are owned by Scripps Networks, could be identified as independent networks. Other cable networks identified as independent networks in other studies, such as the Hallmark Channel and the NFL Network, did not fall into the top 20 cable networks by subscribership in 2008 or in previous years, so they were not included in our analysis.

Combining ownership in both prime time broadcast programming and widely distributed basic cable networks, the major broadcasters have had an interest in a significant share of television programming over the last decade. Independent producers have been a source for a smaller share of prime time broadcast programming. Cable operators without a major broadcaster affiliation are not a source of prime time broadcast network programming,[32] and over the last decade their interest in the top 20 most widely distributed basic cable networks has decreased. However, they make programming decisions for the cable networks they own and determine which cable networks will be carried on their cable distribution systems. FCC annually reports on cable network programming variety and ownership as part of its video competition report, but the report does not assess the extent to which the sources of programming affect variety in television and selection choices for the public.

STAKEHOLDERS CITED ECONOMIC FACTORS, TECHNICAL ISSUES, AND LEGAL CONDITIONS AS AFFECTING THE AVAILABILITY OF INDEPENDENT TELEVISION PROGRAMMING

In Broadcast Television, Economic Factors Influence the Availability of Independent Programming

Industry stakeholders we interviewed stated that the high cost of developing, producing, and distributing television programs is a significant factor that affects the availability of independent programming in broadcast television. According to television broadcast executives and representatives of independent producers, developing and producing broadcast television programs is costly and financially risky. For example, one report estimated that major broadcasters spent about $120 million for the 1997-1998 season to develop 49 drama pilots and used 14 in their schedules, of which 1 program returned for a second season.[33] Moreover, according to television broadcast executives, once programming is developed, the costs to produce a scripted drama or comedy program range from about $21 million to $48 million for 21 program episodes per season, with no guarantee that a program will continue to be produced for another season.[34]

Producers need to sell their program ideas to major broadcasters and secure financing to cover the costs of developing and producing scripted television programs. Because of their large size and access to capital, major broadcaster-affiliated studios and other large unaffiliated studios often have the ability to finance development and production costs.[35] However, representatives of independent producers stressed that it is difficult for them to obtain financing for development and production costs, and oftentimes they must secure financing through the major broadcaster-affiliated studios. The independent producers said since major broadcasters have the ability to finance production costs and make programming decisions, it results in seven or eight companies controlling a significant portion of the program content on television.

When selecting programming for prime time, television broadcast executives told us that they strive to air programming that will achieve high ratings. Advertisers will generally pay more for programs that achieve higher ratings, and since major broadcasters rely on advertising revenue, it is in their financial interest to select programs that will accrue the high level of audience that drives advertising revenue. Television broadcast executives and an academic expert we contacted stated that they also consider quality for prime time programming, and not necessarily the source of programming (i.e., whether the program was produced by an independent producer or an affiliated production studio).[36] They said quality programming will attract the largest share of viewers, which in turn, drives advertising revenue. Further, they stated that since advertisers spend less overall during times of economic downturn and have multiple choices for their advertising dollars (such as on cable television and the Internet), it is all the more essential to have quality programming to attract the advertisers.

While television broadcast executives said that it is the quality, not the source, of programming that influences the selection of prime time programming, major broadcasters are, nevertheless, financially invested in the affiliate-produced programs and stand to gain additional profits if the affiliated programming makes it to syndication. Consequently, some stakeholders said broadcasters might choose their own programming over that of independent producers. In particular, according to an academic expert and representatives of independent producers, if both major broadcaster-affiliated studios and an independent producer offer similar genre and programming content to a major broadcaster, the major broadcaster will select the program from its affiliated studio over an independent producer because of these financial interests. As we previously noted, major broadcaster-affiliated studios (5 companies) produced 82 percent of prime time programming in the fall 2009 prime time schedule. While independent producers most likely would be unable to produce and distribute programming without some financial arrangements with major broadcasters, they said working under the major broadcasters' control could cause them to lose creative control of the program's content, with the writing of the program being directed by the studio bearing the financial risk of production. For example, an independent producer cited the replacement of a writer for CBS's *The Education of Max Bickford,* a drama on the major broadcaster's 2001 prime time schedule, when creative differences arose with the major broadcaster that owned the program.

In Cable Television, Economic Factors, Finite Capacity, and Federal Law Affect Network Carriage

For carriage on cable television, stakeholders cited (1) economic factors, (2) finite capacity, and (3) federal law as affecting carriage of new independent networks.

Economic factors. Representatives of independent networks and some video providers said economic factors affect carriage of new independent networks and their programming. According to video providers, it is difficult to determine the cost and value of new independent networks and how many subscribers will be gained based on concepts and business plans of unproven independent networks. Representatives of independent networks we contacted and a study we reviewed indicated that a new network usually faces considerable uncertainty as to whether it will be distributed by a sufficient number of video providers to make its operations viable. Similarly, an academic study indicates that for new networks, there is a high cost to sustaining operations while attracting a sufficient number of video providers and their subscribers.[37] For instance, one report stated that cable network Fox News Network had invested over $150 million by the time it launched in 1996, but it was expected to lose up to $400 million in the next 5 years.[38] Representatives of independent networks told us that it is difficult to obtain financing for a new cable network because commercial banks want a network to secure carriage with a major cable company, such as Comcast, before extending financing to it.

By contrast, cable networks developed by cable operators, major broadcasters, or other media companies are generally more able to finance the development of affiliated networks over new independent networks. As our analysis indicated, major broadcasters and their affiliated companies owned at least half of the most widely distributed cable networks. Basic cable networks that are affiliated with cable operators, major broadcasters, or other media companies can negotiate carriage of an affiliated cable network as part of an agreement for carriage of an established affiliated network. For example, the Walt Disney Company owns ESPN, SoapNet, and ABC Family cable networks, along with ABC. According to representatives of small cable operators, during the course of negotiating for carriage for ESPN, they must also carry ESPN's spin-off cable networks, including ESPN2 and ESPNEWS. In another example, a new cable network—Wedding Central—that is affiliated with cable operator Cablevision was launched in August 2009 on its distribution system.

Finite capacity. Stakeholders also cited finite capacity in cable system infrastructure of some video providers as a technical issue that affects selection and availability of independent programming. Representatives of video providers we contacted commented that although their overall capacity to carry television programs has expanded with advanced technology, it remains finite. Because cable operators and telecommunications companies offer a wide array of services over their broadband networks, they must determine how to allocate their systems' capacity among these multiple services.[39] Representatives of cable operators and television broadcast executives told us that adding another cable network—independently produced or otherwise—when more than 75 already exist in basic cable, might not be considered the most efficient use of cable operators' resources and capacity. For example, given the demand for high-speed Internet services, cable operators told us they want to ensure they are using the finite capacity of their systems efficiently to be able to meet that demand.

Despite the constraints on capacity in the cable system infrastructure, representatives of video providers and television broadcast executives we spoke with noted that alternative distribution platforms, such as online video streams, have provided more outlets and opportunities for independent programming. For instance, in 2007, two independent producers produced a television drama called *Quarterlife*, which was aired on the social network Web site MySpace.com. On the other hand, television broadcast executives and representatives of independent producers we contacted commented that although the Internet provides the opportunity for distribution of independent programming, it does not translate to success with regard to attracting the number of viewers that television offers.

Federal law. Stakeholders cited, and studies have reported, that FCC rules and regulations implementing certain federal statutes can also influence programming decisions.

- **Retransmission issues.**[40] As we previously mentioned, representatives of some video providers stated that the business practice of bundling networks—meaning that certain networks are sold as a package with broadcast networks rather than being sold individually—which may occur during negotiations between broadcasters (which can include major networks and local stations) and video providers for retransmission rights. Such bundling influences video providers' carriage decisions and limits their ability to select independent programming. In 2004, we reported that because the terms of retransmission agreements often include the carriage of major broadcaster-owned cable networks, cable operators sometimes carry cable networks they otherwise might not have carried.[41]

Representatives of some video providers told us recently that this practice also fills their systems' capacity, leaving less capacity for independent cable networks and making it difficult for independent cable networks to gain carriage. Television broadcast executives, on the other hand, commented that negotiations in lieu of invoking the retransmission rule may be necessary for them to be fully compensated for their content.

As part of its annual report on the status of competition in the delivery of video programming, FCC is currently seeking data and analysis on implementation of the retransmission consent rules.[42] FCC also has a separate proceeding specifically looking at revisions to the retransmission consent rules and whether it would be appropriate to preclude the practice of programmers tying desired programming with undesired programming, such as tying carriage of a major broadcaster-owned cable network to retransmission conditions for a broadcast signal.[43] The comment period for the notice closed in December 2007, and FCC officials are currently reviewing comments.

- **Program carriage rule.**[44] Representatives of independent cable networks and public interest groups stated that although the program carriage rule is needed to promote independent programming, FCC criteria for determining discrimination on the basis of affiliation are unclear. They told us more precise standards for proving discriminatory or exclusionary conduct by cable operators as well as the establishment of a time frame for FCC to determine whether the complaining independent cable networks have sufficient evidence to proceed to a hearing would make the rule more effective. According to independent cable network representatives, some independent cable networks have waited over a year before FCC determined whether it would conduct a hearing. Because the independent cable network is not being carried by the defendant cable operator in the interim, some independent cable networks can go out of business before a decision is made.[45] Representatives of cable operators, on the other hand, stated that the rule is not necessary because a cable operator's decision to reject a network could be based on the program quality and similarity of content and not on the ownership of a network.

- **Leased access rule.**[46] In the case of the leased access rule, a public interest group official indicated that this rule has not achieved what it was intended to do because the prices for leased access were set too high.[47] Representatives of cable operators explained that the rule forces cable operators to carry programming even if they believe the channel does not bring much value to the subscribers. Representatives of cable operators cited home shopping channels as an example of programming that relies on leased access to gain carriage. The rule also affects the cable operators' ability to carry other programming because the set-aside channels consume capacity that could be used for other programming. Cable operators also noted that the rule does not apply to satellite providers and their systems.

STAKEHOLDERS CITED VARIOUS FACTORS AS INFLUENCING PROGRAMMING DECISIONS, INCLUDING FORMAT AND PLAYLIST SELECTION, IN COMMERCIAL AND PUBLIC RADIO

Commercial Radio Stations Largely Make Programming Decisions Based on Economic Factors

In selecting radio station formats and music playlists, stakeholders we interviewed stated that (1) advertisement revenue, (2) cost of programming, and (3) market competition are key economic factors that influence programming decisions in commercial radio.

Advertisement revenue. Commercial radio stations are primarily funded by advertisement revenue obtained from selling radio time to companies seeking to reach specific demographic segments.[48] Radio station owners and experts told us that when making decisions about format and playlist selection, program directors will consider the number of listeners that programming will likely attract,[49] and, in turn, the advertisement revenue they may earn.[50] The rates that a station obtains for advertising time depend on the station's ability to attract listeners within the advertisement companies' target demographic segment, the length of the advertisement spot, and the size of the market, with larger markets typically receiving higher rates than smaller markets. Radio stations compete for listeners and advertising revenue with other stations within their respective local markets. Consequently, radio stations continuously examine their programming content to try to attract an audience that is highly desirable to advertisers. In particular, a radio station's format enables it to target specific segments of listeners sharing demographics that appeal to advertisers. According to a radio industry expert, if the advertising market is not interested in reaching the specific target audience of a music format, the station will not be able to survive economically because it will not be able to gain enough ad revenue. Moreover, radio station owners with stations in different markets but of the same format can be more effective at attracting revenue from advertisers who want to reach a similar demographic in multiple markets.

Cost of programming. Another economic factor that influences programming decisions is the cost to produce radio content. For example, radio station owners and experts told us that increased costs and decreased advertisement revenue over the past decade have led to an increase in the use of voice tracking and syndicated programming.[51] According to radio station owners and experts, voice tracking is less costly than producing shows for individual markets, and to save programming costs, some stations choose to import programming from another market during peak listener times rather than hire their own radio personalities. In addition, radio industry experts pointed out that historically, stations in small markets have generally relied on nationally syndicated programming to bring in marketable talent that will allow them to compete with other stations in the market. Some stakeholders have expressed concern that voice tracking and syndicated programming are replacing local programming and therefore the needs and interests of the local community are not being reflected by the voice-tracked or syndicated programming. However, representatives of radio station owners have stated that there is no evidence that voice tracking or syndicated programming diminishes localism. For example, one station owner pointed out that the value of programming is determined by how strongly it resonates with listeners, regardless of where it originates.

Number of stations

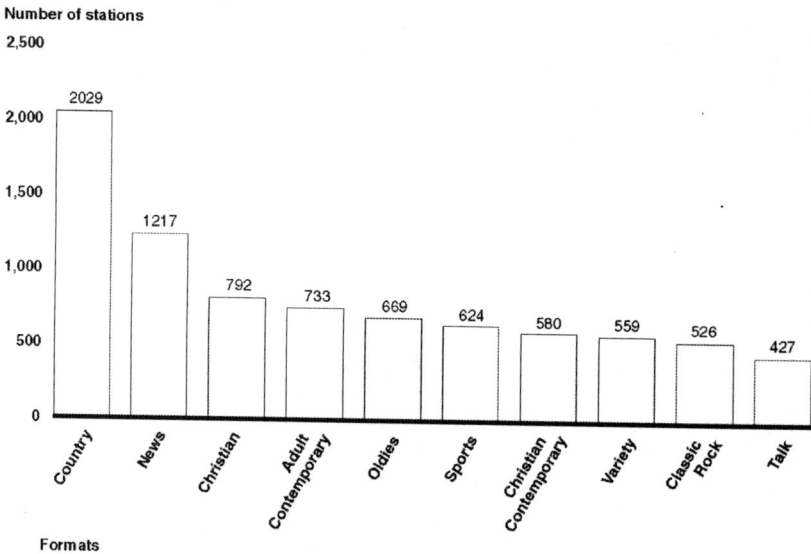

Source: GAO analysis of BIA*fn* data.

Note: The "Primary Format" field in the BIA*fn* database was used to identity the most popular formats in 2009 and includes data for 14,628 stations, both in and outside of Arbitron-rated markets, and commercial and public stations. Stations designated as multicast stations, or having a construction permit, are not included in this analysis because multicast stations are available only to listeners who have purchased an HD radio and radio stations designated as having construction permits are not yet operating.

Figure 6. Most Popular Formats Nationwide in 2009

Market competition. Marketplace factors, such as the extent of competition in a given market, also affect programming decisions. For example, radio station owners stated that when radio station program directors are trying to determine a station's format, they will consider what formats are currently available in the local market and what formats are missing. If there are already stations programmed with a popular format in a given market, a radio station will likely look to competitively differentiate itself by selecting a format targeted toward a demographic that is not currently being served. In doing so, a station may also better compete for audiences and advertising revenues with other media.

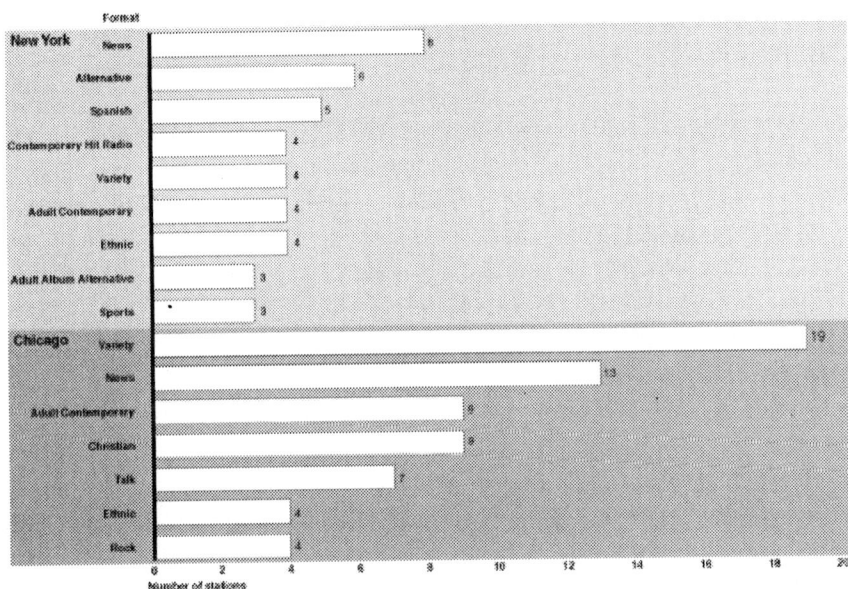

Source: GAO analysis of BIAfn database.
Note: Datasets for Arbitron Market 1 (New York) included data for 74 radio stations
and 130 radio stations for Arbitron Market 3 (Chicago). Both datasets included
commercial and public FM and AM stations.

Figure 7. Most Popular Formats in Arbitron Market 1 (New York) and Arbitron
Market 3 (Chicago) in 2009

Experts and representatives of independent producers told us that radio
station formats have become more specific in recent years in an attempt to
enable stations to target a specific demographic and attract advertisers, and as
a result, radio station formats have changed over time. According to radio
station owners, the number of radio station formats has increased.
Representatives of radio station owners conducted a study examining radio
station formats, and found that from 2001 to 2005, the number of radio station
formats increased by 7.5 percent. Station owners have characterized this
increase in the number of formats as an increase in variety in radio
programming. However, some experts and representatives of independent
producers have noted that formats with different names often have similar
playlists, diminishing real variety among those formats. For example, one
expert noted that it is very difficult to discern differences in playlists between
radio formats such as Rock and Light Rock.

Local Interests also Affect Programming Decisions in Both Commercial and Public Radio Stations

Stakeholders stated that in both commercial and public radio, programming decisions such as selection of format and music playlists are based on the interests of listeners in a given market. Radio station owners in both commercial and public radio reported that program directors will conduct research related to the demographics and preferences of the listeners in their markets to ensure they are meeting the needs of their community. In commercial radio, understanding the interests of listeners in a given market is important for the station to attract a large audience and, as previously noted, attract advertising revenue. According to radio station owners, program directors are expected to be familiar with music interests in their markets and make programming decisions that will be successful in reaching an audience within their market. A stakeholder also noted that even among similarly formatted radio stations, the playlist will vary to meet the needs of the local market. For example, the type of country music that is popular in Tucson, Arizona, can be very different from popular country music in New York City.

According to our analysis, in 2009, the 10 most common formats across all national radio stations included Country, News, Christian, Adult Contemporary, Oldies, Sports, Christian Contemporary, Variety, Classic Rock, and Talk, as shown in figure 6.

We found that within selected individual markets, the top radio formats differ from the top radio formats nationally, indicating that programming decisions are locally based on the preferences and interests of listeners within a given market. For example, the most popular radio station formats in New York City (the largest Arbitron market) include 5 formats not reflected in the top 10 national radio formats (Alternative, Spanish, Contemporary Hit Radio, Ethnic, and Adult Album Alternative). In addition, we found 19 percent of all stations in the New York market were designated as Ethnic and Spanish formats compared with 7 percent nationwide, suggesting that programming decisions among radio stations in this market reflect the demographics and interests in the market.[52] By comparison, in Chicago, Illinois (the third-largest Arbitron market), we found that 11 percent of stations in this market were designated as Ethnic and Spanish formats. Furthermore, formats that were among the most popular in Chicago but not in New York included Christian, Talk, and Rock (see figure 7).

As is the case in commercial radio, representatives of public radio reported that programming decisions are locally based on the preferences and interests of listeners within a given market; however, they said their community service orientation also influences programming decisions. Representatives of public radio explained that local public stations select their own formats and determine their own audience strategies based on their understanding of local community needs, and their role in serving those needs. They also said the cost of programming is a final consideration for public radio stations after quality- and mission-related factors are considered. In addition, representatives of public radio noted that public stations generally play music from artists that are signed to small, independent labels.[53] Independent labels generally seek out a station if the station's format includes music similar to that of the labels, and will then establish relationships with such stations. On the basis of our review of 2009 format data for commercial and public radio stations, we found that the top 10 formats in public radio differ from the top 10 formats in commercial radio (see figure 8). Only two formats (News and Spanish) were among the top 10 formats in both commercial and public radio.

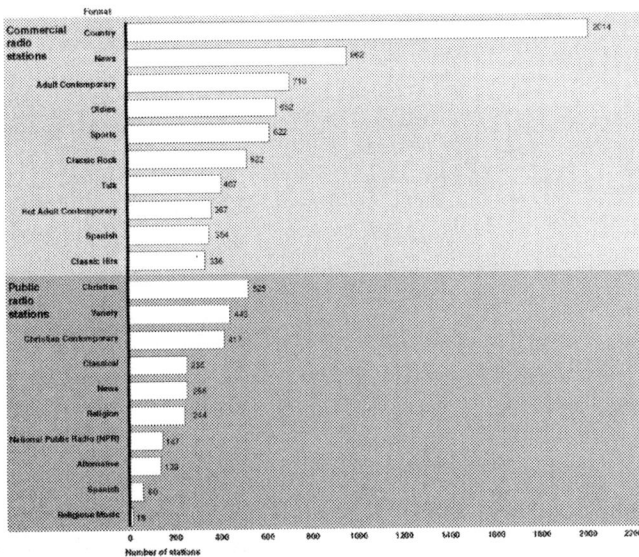

Source: GAO analysis of BIA*fn* data

Note: Data used for analysis include 6,946 commercial stations and 2,503 public stations, both inside and outside of Arbitron-rated markets, and FM and AM stations. Stations designated as multicast stations, or having a construction permit, are not included in this analysis.

Figure 8. Top 10 Formats among Commercial and Public Radio Stations

Opinions Vary on How Consolidation in the Radio Industry Has Affected Programming Decisions

Stakeholders that we interviewed generally agreed that since 1996, the number of stations owned by a single radio station owner has increased; however, viewpoints varied about the extent to which consolidation has affected programming decisions. Experts and representatives of independent producers we contacted stated that the elimination of the radio ownership limits in 1996 resulted in an increase in the number of stations owned by a single station owner nationally and in local markets. Independent producers have reported that the radio station holdings of the 10 largest radio station owners have increased significantly. On the basis of our analysis, we found that the share of commercial stations owned by the top 10 station owners did increase, from 4 percent in 1996 to 20 percent in 2009. However, throughout that period, the top 10 radio station owners did not own more than 21 percent of all commercial stations, as shown in figure 9.

Percentage owned

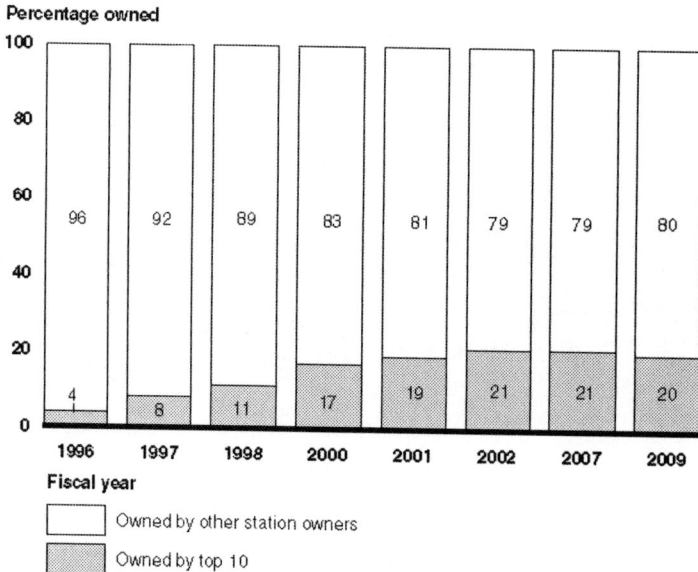

Source: GAO analysis of FCC and BIA*fn* data.
Note: Because of data limitations, we did not include data for 2003 through 2006 and for 2008.

Figure 9. Share of Commercial Radio Stations Owned by Top 10 Radio Station Owners for Selected Years

In addition, we analyzed data for the top 10 national radio station owners in 2009 and found that for most owners (7 out of the 10 owners), stations' formats were differentiated within individual markets. For example, Clear Channel—the largest radio station owner—owns multiple radio stations in 148 Arbitron markets. We found that in most of those markets (72 percent), Clear Channel programmed its stations with different formats, while in 28 percent of those markets some stations were programmed with the same format. As illustrated in table 3, among the station owners that we reviewed, those with the highest percentage of overlap among radio stations in the same market included American Family Association (78 percent), Cox Radio (56 percent), and Educational Media (56 percent). We also found that 75 percent of the markets where format overlap did exist included large markets with 30 or more radio stations.

Table 3. Percentage of Format Overlap among Top 10 Radio Station Owners in 2009

Radio station owner company	Number of Arbitron markets where owner is present	Number of Arbitron markets where owner owns multiple stations	Percentage of markets where owner owns multiple stations with the same format
Clear Channel Communications Inc.	153	148	28
Educational Media Foundation	103	34	56
Cumulus Broadcasting LLC	58	54	19
Citadel Communications	48	43	33
American Family Association Inc.	45	9	78
Salem Communications Corporation	35	28	39
CBS Radio	30	28	21
Entercom	24	21	33
Cox Radio Inc.	19	16	56
Saga Communications Inc.	14	13	15

Source: GAO analysis of BIA*fn* data.

Radio station owners and representatives of independent producers offered different perspectives on how consolidation in the radio industry has affected programming decisions nationally and in individual markets. On one side, radio station owners and experts told us that to remain financially viable, stations have had to eliminate duplicative operating and overhead expenses and establish a business model where one program director is responsible for programming decisions for multiple stations. Some station owners added that program directors overseeing programming decisions for stations in multiple markets make decisions based on the interests of listeners within the individual markets. Further, radio station owners and experts have reported that common ownership of multiple stations in a single market benefits the audience in that market, as the station owner will choose to diversify formats among its stations to attract a large share of the listening audience in the market.[54]

Another viewpoint expressed by representatives of independent producers and experts is that the increased consolidation has changed the stations' decision-making structure, resulting in homogenized programming decisions across markets and resulting in large companies using centralized methods to make programming decisions. According to this view, as jobs are consolidated when one entity owns multiple stations, one program director may make similar programming decisions across multiple stations in different markets. The independent producers said that as a result, playlists of radio stations owned by the same owner will overlap. Studies conducted by representatives of independent producers and academic experts examined playlists of radio stations owned by the same owner across all markets and found overlap in playlists of stations with the same format.[55] For example, a December 2006 study published by the Future of Music Coalition found examples of overlap among playlists of individual stations owned by the same company in different markets—such as an overlap for the playlists of two country stations located in different markets (WQRB-FM in Eau Claire, Wisconsin, and WRWD-FM in Poughkeepsie, New York). However, the study did not examine overlap and differences among playlists of owners' radio stations in the same market. A January 2006 study conducted by an academic expert also examined playlist data for each owner's radio stations and found that the playlists of radio stations in different markets overlapped, but that the playlists of radio stations in the same market were different.

APPENDIX I. OBJECTIVES, SCOPE, AND METHODOLOGY

To obtain information on the extent to which sources of programming in television have changed over last decade, we analyzed available data on two major sources of television programming—companies producing prime time broadcast television programs and companies with cable channel ownership interests during the last decade. We focused on programs broadcasted during prime time because that is the block of time on television with generally the most viewers and in turn generates the most advertisement revenue for networks.[56] To determine which companies produced prime time broadcast television programming, we used a previous Federal Communications Commission (FCC) study and the *International Television & Video Almanac* to classify prime time programs into two categories: (1) programs produced by major broadcasters, and (2) programs produced by independent production companies not affiliated with a major broadcaster (independent producers). We analyzed the fall prime time schedules in 2002, 2005, 2008, and 2009 and classified them in the two categories. We selected these years because annual data that tracked production information in the two categories were limited; FCC's previous study contained data in the two categories for 2002. We then conducted our analysis for every third year (2005 and 2008) using the *Almanac's* production company information for each television program in that year's debut fall broadcast prime time schedule and classified the programs into the two categories. We also analyzed the *Almanac* for the 2009 fall prime time schedule to provide the most current data available. Additionally, since basic cable networks are also a source of television programming, we analyzed the ownership of those networks as an indicator of which entities control the television programming on the networks. To determine cable network ownership over the last decade, we used data from SNL Kagan, which show companies having an ownership interest in each of the basic cable networks from 1998 to 2008.[57] We analyzed these data to determine the types and the number of companies that have had an ownership interest in basic cable networks and the companies that owned the largest number of networks during this period. To analyze cable network ownership for the most widely distributed networks, we used the 20 basic cable networks with the most subscribers (the top 20 networks) from 1998 to 2008 and classified the networks into one of four categories: (1) networks owned by major broadcasters, (2) networks owned by video providers, (3) networks owned by both major broadcasters and video providers, and (4) networks owned by other types of companies. We also examined the top 20 networks in 2008 for any independent cable networks; that is, any network that did not have an affiliation with a major broadcaster or video provider, or an affiliation with a major holding company with media interests.

Table 4. Experts and Industry Stakeholders We Contacted

Stakeholder groups	Stakeholder
Academic experts	Mara Einstein, New York University and Queens College
	David Waterman, Indiana University
	Philip Napoli, Fordham University
	Andrew Sweeting, Duke University
Broadcast television and affiliates groups	The Walt Disney Company/ABC
	CBS Corporation
	NBC Universal
	News Corporation/Fox
	CBS and NBC Affiliate Association
	Sinclair Broadcast Group
	Young Broadcasting
Radio station owners group	Clear Channel Communications
	Citadel Broadcasting
	American Public Media/Minnesota Public Radio
Public television and radio	Association of Public Television Stations
	Public Broadcasting Service
	National Federation of Community Broadcasters
	National Public Radio
Video providers	Comcast
	DirecTV
	DISH
	AT&T
	Verizon
Cable networks	Discovery Communications
Industry associations	American Cable Association
	National Association of Broadcasters
	National Association of Independent Networks
	National Cable and Telecommunications Association
Independent programming groups	Independent Film and Television Alliance
	Writers Guild of America, West
	Center for Creative Voices in Media
	American Association of Independent Music
	Future of Music Coalition
Public interest, nonprofit groups	Consumers Union
	Consumer Federation of America
	Media Access Project
	Progress and Freedom Foundation
	Free Press
Industry Research Group	Edison Research

Source: GAO.

To determine the factors and conditions that stakeholders identified as affecting the availability of independent programming in television and factors that influence radio programming decisions, we interviewed or obtained written comments from a variety of experts and industry stakeholders, including academics, industry representatives, media companies, and public interest groups (as shown in table 4) to obtain their views on the factors that affect the availability of independent programming in television and radio.

We selected the experts and stakeholders based on relevant published literature, including FCC filings and reports, stakeholders' recognition and affiliation with a segment of the media industry (i.e., cable operators, satellite providers, broadcasters, radio station owners, independent radio advocacy groups, and so forth), and other stakeholders' recommendations. In our selection of experts and stakeholders, we intended to obtain balanced and diverse views; we did not weight experts' and stakeholders' views but grouped similar stakeholders that represent a segment of the media industry. We conducted semistructured interviews and analyzed the responses to determine patterns and the extent to which the experts and stakeholders agreed on the key factors affecting independent programming and radio programming decisions. We also spoke with FCC officials and reviewed the relevant laws, regulations, literature, comments filed by stakeholders in various FCC proceedings, FCC studies, and FCC-sponsored research on television and radio programming.

In addition, for radio, we examined radio station formats, indicating the genre and types of programming, such as Adult Contemporary, Country, News, Sports, and Talk, a station might play. We obtained historical data on the distribution of radio stations by their primary formats nationwide and in local markets from 1999 to 2003 and format data from the Broadcast Investment Analyst Financial Network's (BIA*fn*) Media Access Pro Database, containing station-level data for commercial and public radio stations in the United States from 2004 to 2009. Although the BIA*fn* format data provide a general overview of the genre of programming aired on a given radio station, they do not identify specific programming content that is played on the station. We did not look at independently produced programming on radio because national playlist data identifying record label affiliation are not available. We analyzed the data to determine programming variety and distribution of radio stations by their format nationwide and in local markets in 2009. To highlight programming variety in local markets, we selected two radio station markets-New York and Chicago-and analyzed the format data of radio stations in those markets and compared them with national radio station format data in 2009. We selected New York and Chicago because these two markets are similar in size, (New York is the largest market, and Chicago is the third-largest market) but have different demographic populations. In addition, each market contains both commercial and public stations, FM and AM stations, and contains multiple radio station owners in the market. To highlight similarities and differences in programming variety among commercial and public stations, we examined 2009 format data for commercial and public radio stations and identified the top 10 most popular formats (based on the number of stations with the particular formats available) for each group nationwide. Finally, to examine programming variety for each owner's radio stations and consolidation in the radio industry, we selected the top 10 radio station owners—that is, owners who own the most radio stations nationwide—and reviewed format data of stations owned by the top 10 owners. To identify the top 10 radio station owners in 1996-1998, 2000-2002, 2007, and 2009, we used data from FCC reports and the BIA*fn* database. The top 10 radio station ownership data were not available in 2003-2006 and 2008. Collectively, in 2009, the top 10 owners owned a total of 2,262 commercial radio stations, or 20 percent of all U.S. commercial radio stations. In addition, the top 10 owners owned stations that reach a 44 percent share of total Arbitron listeners in the United States and collect 52 percent of the radio industry's revenue. For each station owner, we then examined similarities and differences in formats among commonly owned radio stations in the same market. We also reviewed studies on radio programming for information on radio station playlists and the extent to which playlists for commonly owned radio stations overlap in the same market.

To assess the reliability of the basic cable network data obtained from SNL Kagan, and radio data obtained from BIA*fn* used in our analysis, we (1) obtained information from the system owners on their data reliability procedures, (2) reviewed systems documentation, (3) reviewed data to identify obvious errors in accuracy and completeness, and (4) compared the data with information we obtained from other sources, including FCC studies. After reviewing the data sources, we determined that the data were sufficiently reliable for the purposes for which we have used them in this report.

We conducted our work from May 2009 to March 2010 in accordance with all sections of GAO's Quality Assurance Framework that are relevant to our objectives. The framework requires that we plan and perform the engagement to obtain sufficient and appropriate evidence to meet our stated objectives and discuss any limitations in our work. We believe that the information and data obtained, and the analysis conducted, provide a reasonable basis for any findings and conclusions in this product.

APPENDIX II. REGISTERED MARKS USED IN THE REPORT

Registered mark	Listed owner
20th Century Fox	Twentieth Century Fox Film Corporation
ABC	American Broadcasting Companies Inc.
ABC Family	American Broadcasting Companies Inc.
Arbitron	Arbitron Inc.
AT&T	AT&T Intellectual Property II L.P.
BIAfn	BIA Financial Network Inc
Cablevision	Cablevision Systems Corporation
CapStar Broadcasting	CapStar LLC
CBS	CBS Broadcasting Inc.
CBS Radio	CBS Broadcasting Inc.
Citadel Communications Company	Citadel Communications Company
Clear Channel	Clear Channel Communications Inc.
CNN	Cable News Network LP
Comcast	Comcast Corporation
Comcast SportsNet	Comcast Sports Management Services LLC
Consumers Union	Consumers Union of United States Inc.
Consumers Federation of America	Consumers Federation of America
Cox Communications	Cox Communications Inc.
Cox Radio	Cox Radio Inc.
DirecTV	Directv Inc.
Discovery Channel	Discovery Communications LLC
DISH Network	DISH Network LLC
Disney Channel	Walt Disney Productions Corporation
Edison Research	Edison Media Research Inc.
Entercom	Entertainment Communications Inc.
ESPN	ESPN Inc.
Fox	Twentieth Century Fox Film Corporation
Fox News	Twentieth Century Fox Film Corporation
GE	General Electric Company
Hallmark	Hallmark Licensing Inc.
HBO	Home Box Office Inc.
HGTV	Scripps Networks Inc.
Independent Film and Television Alliance	Independent Film & Television Alliance Corporation

(Continued)

Registered mark	Listed owner
ION Television	ION Media Networks Inc.
Lifetime Networks	Lifetime Entertainment Services LLC
Media Access Project	Media Access Project
Minnesota Public Radio	Minnesota Public Radio
MTV Networks	Viacom International Inc.
MyNetworkTV	MynetworkTV Inc.
National Public Radio	National Public Radio, Inc.
NBC	NBC Universal, Inc.
News Corporation	News Holdings Pty Ltd.
NFL Network	National Football League
Paramount Television	Paramount Pictures Corporation
Saga Communications Inc	Saga Communications Inc.
Salem Communications Corporations	Salem Communications Corporation
Scripps Networks	Scripps Howard Broadcasting Company
Showtime	Showtime Network Inc.
SNL	SNL Financial LC
SoapNet	Disney Enterprises Inc.
Sony Pictures	Sony Corporation
Starz	Starz Entertainment Group LLC
TBS	Superstation Inc.
The Food Network	Television Food Network, G.P. , et Al.
Time Warner Cable	Time Warner Inc.

Source: Trademark Electronic Search System, United States Patent and Trademark Office.

APPENDIX III. COMMENTS FROM THE FEDERAL COMMUNICATIONS COMMISSION

Federal Communications Commission
Washington, D.C. 20554

March 1, 2010

David Wise
Director, Physical Infrastructure Issues
United States Government Accountability Office
Washington, D.C. 20548

Re: GAO-10-369

Dear Mr. Wise:

Thank you for the opportunity to review and comment on the Government Accountability Office Draft Report *Media Programming – Factors Influencing the Availability of Independent Programming in Television and Programming Decisions in Radio.*

While the Draft Report does not contain any specific recommendations for action by the Federal Communications Commission, the Draft Report does recognize the important role that the media industry plays in educating and entertaining the public and acknowledges that the Commission's rules can affect programming decisions. To that end, the Commission's longstanding goals in the development and implementation of media policy have been the promotion of diversity, competition, and localism in media. The report covers several subjects pertinent to our media ownership policy. The Commission does not propose any editorial recommendations to the Draft Report. In an attachment to this letter, however, we offer technical corrections for your consideration.

As the Draft Report notes, the Commission is required to review its media ownership rules every four years to determine whether those rules are necessary in the public interest. Currently, the Commission is in the early stages of the 2010 quadrennial media ownership review. The Draft Report's availability comes at a time that its information and findings can assist the Commission as it moves forward in the ownership review.

Thank you for the opportunity to comment on the Draft Report. The FCC appreciates your contribution regarding the important subject areas covered.

Sincerely,

William T. Lake
Chief, Media Bureau

ATTACHMENT

On p.1, clarify the actions taken by the Telecommunications Act of 1996 ("The 1996 Act"). The 1996 Act eliminated nationwide broadcast television and radio station ownership limits and relaxed (rather than "reduced") local broadcast radio station limits. The 1996 Act did not change local television ownership limits, but directed the FCC to conduct a rulemaking to evaluate its existing local television ownership limitations. See 1996 Act § 202.

Regarding p.8 n.15, at the time of the 1996 Act's enactment, the FCC's rules contained a national radio limit of 20 AM plus 20 FM stations and a local radio limit of two or three AM plus two or three FM stations, depending on the size of the local market. See Revision of Radio Rules and Policies, 9 FCC Rcd 7183 (1994).

Regarding the second paragraph of the "Retransmission Consent" bullet on p.23-24, the FCC released an Order on January 10, 2010 in the program access proceeding adopting rules permitting complainants to pursue program access claims involving terrestrially delivered, cable-affiliated programming similar to the claims that they may pursue for satellite-delivered, cable-affiliated programming, where the purpose or effect of the challenged act is to significantly hinder or prevent the complainant from providing satellite cable programming or satellite broadcast programming. See 2009 WL 236800 (released Jan. 20, 2010).

End Notes

[1] Independently produced programming has no affiliation with broadcast networks, cable operators, or satellite providers.

[2] *In re Review of the Syndication and Financial Interest Rules, Sections 73.659 –73.663 of the Commission's Rules,* F.C.C. 95-382, 10 F.C.C.R. 12165, 10 F.C.C. Rcd. 12165, (1995).

[3] Pub. L. No. 104-104, Title II, § 202, 110 Stat. 110, 111-112, as amended by Pub. L. No. 108-199, Div. B, Title VI, § 629, 118 Stat. 3, 99-100 (2004) (47 U.S.C. § 303 Note).

[4] A broadcast network is an organization that may produce and distribute television programs, such as drama, comedy, reality programs, and news, to the public through local television stations. We refer to the broadcast networks (ABC, CBS, Fox, and NBC) as major broadcasters in this report, although the term "broadcasters" can include local television stations that are not owned by the major broadcast networks. Throughout this report, we refer to media firms by their popularly known acronyms or names, some of which like ABC, CBS, Fox, and NBC, are registered trade or service marks. Appendix II contains a list of registered marks appearing in this report.

[5] A basic cable network is an organization that may produce television programs, which are distributed to the public through a subscription video service.

[6] To analyze the fall prime time schedule in each year, we included programs on the schedule Monday through Saturday from 8 p.m. to 11 p.m., and on Sunday from 7 p.m. to 11 p.m. Because some prime time program schedule changes or cancellations can occur in the fall prime time schedule, we used the debut schedule of programs that appeared in September of each selected year.

[7] SNL Kagan Media and Communications is a private research company that collects and maintains data on cable networks.

[8] A radio station format refers to the type of programming content on a radio station, such as Adult Contemporary, Country, Jazz, News and Information, Sports, Talk, and so forth.

[9] Local markets, also known as Arbitron markets, are geographically contiguous areas in which Arbitron Inc. surveys the listenership of radio stations for rating. These markets align with the Office of Management and Budget's Metropolitan Statistical Area (MSA) definitions. As of August 2009, there were 300 Arbitron markets.

[10] Public radio stations are locally owned and operated stations that receive some or all of their funding from listener contributions, the federal government, or other sources. Some public

radio stations are affiliated with National Public Radio, which is a national radio service that provides station programming content.

[11] Affiliated stations are stations not owned by major broadcasters but grant broadcasters use of specific time periods for network programming and advertisement, for compensation.

[12] Video providers offer subscribers television programming through different service tiers—that is, bundles of networks grouped into a package. A basic tier, which is the lowest level of cable services, includes the local broadcast stations. Video providers also offer an expanded basic service tier, which expands upon the basic service. Additionally, subscribers can also purchase digital tiers and premium pay channels, such as HBO and Showtime, for an additional fee.

[13] Video providers must also pay licensee fees—usually on a per subscriber basis—for the rights to carry cable networks and their respective programming.

[14] The number of stations includes full-service commercial and public stations and excludes low-power FM stations and stations with construction permits.

[15] Pub. L. No. 104-104, §202(h), 110 Stat. 56. In response to the directive in the 1996 Telecommunications Act, FCC eliminated the nationwide radio ownership limits rule.

[16] In 1999, CapStar Broadcasting and Chancellor Media Corporation merged and were later acquired by Clear Channel Communications Inc., which subsequently became the largest radio station owner, with over 1,000 radio stations in 2007. In 2006, Clear Channel sold off several of its radio stations.

[17] In re Review of the Syndication and Financial Interest Rules, supra. The Fin-Syn rules were adopted in 1970 to limit broadcasters' control over television programming and restricted their ability to own and syndicate programming.

[18] Broadcasters can request syndication rights for programs when they are negotiating the financing of programs. If a program remains on the network long enough to accumulate about 80 to 100 episodes, then the program can be syndicated, that is, offered and sold to local television stations or cable networks for the right to broadcast the programs again during non-prime time hours.

[19] Telecommunications Act of 1996, § 202, as amended (47 U.S.C. § 303 Note).

[20] The Cable Televisions Consumer Protection and Competition Act of 1992, Pub. L. No. 102-385, 106 Stat 1460, established these rules, placed in Sections 325, 614, and 616 of the Communications Act, as amended (47 U.S.C. § 534). FCC then established regulations to put into effect those statutory provisions. (In addition, the act established the program access rule to prevent a vertically integrated cable operator from discriminating in the process, terms, and conditions that it makes programming available to unaffiliated distributors or have exclusive access to the programming in which it has an ownership interest. Because stakeholders we interviewed did not identify how this rule affects independent programming, we did not include a discussion of the issue.)

[21] Communications Act of 1934, § 614, as amended (47 U.S.C. § 534); 47 C.F.R. §§ 76.56, 76.64.

[22] Compensation can take the form of cash payments, the video provider's purchase of advertising time on the broadcast station, the broadcaster being given free advertising time on the video provider's system, the video provider's carriage (and tier placement) of other program networks owned by the broadcaster, or some combination of these.

[23] Communications Act of 1934, § 616, as amended (47 U.S.C. § 536); 47 C.F.R. §§ 76.1301, 76.1302.

[24] Communications Act of 1934, § 612, as amended (47 U.S.C. § 532); 47 C.F.R. §§ 76.970, 76.971.

[25] Cable operators with 36 to 54 activated channels must set aside 10 percent, while cable operators with 55 to 100 activated channels must set aside 15 percent of those channels not otherwise required for use or prohibited from use by federal law or regulation. Cable operators with more than 100 activated channels must designate 15 percent of such channels for commercial use.

[26] The Cable Television Consumer Protection and Competition Act of 1992, Pub. L. No. 102-385, § 9, 106 Stat. 1460 (codified at 47 U.S.C. § 532)

[27] Warner Bros. is affiliated with the CW Television Network.

[28] Mara Einstein, "Program Diversity and the Program Selection Process on Broadcast Network Television," *Federal Communications Commission Media Ownership Working Group*, Washington, D.C., September 2002.

[29] Our analysis included ownership interests in basic cable networks, which are those cable networks that often appear in the basic service tier for consumers, and did not include other cable networks carried on a digital tier and premium cable networks, such as HBO, Showtime, and Starz.

[30] During the last decade, the number of basic cable networks has ranged from about 160 to 180.

[31] Because of data availability, our analysis was for the 11 years from 1998 to 2008. In 2009, Time Warner announced a spin-off of its Time Warner Cable operations.

[32] In December 2009, Comcast, a cable operator, announced that it had signed a definitive agreement with GE (owner of NBC) to form a joint programming venture that will be 51 percent owned by Comcast, 49 percent owned by GE, and managed by Comcast.

[33] Joe Schlosser, "Wolf Says Shows Can Fly without Pilots," *Broadcasting & Cable*, July 6, 1998, 30.

[34] In contrast, nonscripted programs are less costly to produce because there are no costs associated with ordering scripts and pilots.

[35] An example of a large unaffiliated studio is Sony Pictures Television Studio, which is not affiliated with any broadcasters. Although large unaffiliated studios may have the ability to finance production costs, stakeholders said they produced fewer programs than broadcaster-affiliated studios because one of the few unaffiliated studios decided to primarily produce movies rather than television programs.

[36] Although quality is subject to the preferences of the networks that select the programs, according to television broadcast executives, general criteria for quality include good ideas, mass audience appeal, and whether the content of the program meets the specific needs of the network. Literature we reviewed indicated that quality programming depends on talent and high production values, both of which can be costly.

[37] Richard Caves, Karen Guo, Catherine O'Gorman, Matthew S. Rosenberg, and Richard J. Wegener, *Switching Channels: Organization and Change in TV Broadcasting*. (Cambridge, Mass.: Harvard University Press, 2005).

[38] Jim McConville, "Fox Ready to Roll Dice in All-News Gamble," *Broadcasting & Cable*, October 7, 1996, 52.

[39] The array of services that cable operators and telecommunications companies offer includes high-speed Internet, telephone, and digital television, as well as traditional analog television services. A satellite provider we interviewed stated that competing uses for broadband is not an issue, since it cannot offer high-speed Internet and telephones services, but it noted that available capacity on its basic service tier is also limited.

[40] 47 C.F.R. § § 76.56, 76.64. As previously mentioned, the retransmission consent rule requires local broadcast stations, some of which are owned and operated by broadcasters, who have opted against must carry status, to negotiate individual retransmission consent agreements with each cable operator in its service area for compensation in exchange for the cable operator's right to carry the broadcast signal.

[41] 47 C.F.R. § § 76.56, 76.64. As previously mentioned, the retransmission consent rule requires local broadcast stations, some of which are owned and operated by broadcasters, who have opted against must carry status, to negotiate individual retransmission consent agreements with each cable operator in its service area for compensation in exchange for the cable operator's right to carry the broadcast signal.

[42] *In the Matter of the Annual Assessment of the Status of the Competition in the Market for the Delivery of Video Programming*, FCC MB Docket No. 07-269. FCC 07-207, 24 F.C.C.R.

750, 24 FCC Rcd. 750 (January 16, 2009), as supplemented at 24 F.C.C.R. 4401, 24 F.C.C. Rcd. 4401 (April 9, 2009).

[43] *In the Matter of the Review of the Commission's Program Access Rules and Examination of Programming Tying Arrangements,* FCC MB Docket No. 07-198, Notice of Proposed Rulemaking, 22 F.C.C. Rcd. 17791 (October 1, 2007).

[44] 47 C.F.R. §§ 76.1301, 76.1302. As previously noted, the program carriage rule prevents video providers from requiring financial interest in programming as a condition for carriage.

[45] FCC officials told us that it takes time to adjudicate these cases because program carriage disputes are complicated and often result from behavior related to program carriage negotiations and require an evaluation of facts and behavior. They further noted that an action one party views as prohibited under the statute may be viewed by the other as a legitimate business practice. In some instances, additional measures are required to perform a proper evaluation. In the event that the staff is unable to resolve a program carriage complaint on the basis of the written record, a case may be sent for a hearing before an administrative law judge.

[46] 47 C.F.R. §§ 76.970, 76.971. The commercial leased access rule requires cable operators to set aside a certain number of channels that can be leased out to independent cable networks for access on its distribution system.

[47] FCC reviewed the price rate for leased access and announced its decision to reduce the charge in November 2007. *In the Matter of Leased Commercial Access,* FCC 07-208, 23 F.C.C.R. 2909, 23 F.C.C. Rcd. 2909 (Feb. 1, 2008). However, FCC's decision has been appealed.

[48] In contrast, public radio stations are primarily funded by contributions received from listeners and, in some cases, government funding.

[49] By "format" of radio programming, we mean the genre of programming content on a radio station, such as Country, Sports, Adult Contemporary, Smooth Jazz, and Rock.

[50] Advertisers and radio stations use data published by audience measuring services, such as Arbitron Inc., to estimate the number and demographics of listeners within an Arbitron radio market.

[51] Voice tracking occurs when a radio station personality prerecords a program that is then aired in multiple markets, including markets other than that of the local radio station. Syndicated programming includes programming that is purchased by a radio station (such as the Ryan Seacrest Top 40 Program) to air on multiple stations in different markets.

[52] FCC commissioned 2007 Media Ownership Study 5 (Tasneem Chipty, *Station Ownership and Programming in Radio,* Boston, Mass.: CRA International, June 24, 2007) indicated that ethnic formats included Asian, Greek, Hawaiian, International, Japanese, Korean, Polish, and Portuguese formats. Spanish formats included Hurban, Mexican, Ranchera, Reggaeton, Spanish, Spanish Adult Contemporary, Tejano, and Tropical formats.

[53] An independent label refers to a record label that is not associated with a major record label.

[54] A 2007 FCC-commissioned study (Tasneem Chipty, *Station Ownership and Programming in Radio,* Boston, Mass.: CRA International, June 24, 2007) and other academic studies (Andrew Sweeting, *Too Much Rock and Roll? Station Ownership, Programming, and Listenership in the Music Radio Industry,* Evanston, Ill.: Northwestern University, January 15, 2006) found similar results, finding that markets with large radio ownership groups offer more format choices within given markets.

[55] While our analysis examined format data, these studies looked at playlists.

[56] To analyze the fall prime time schedule in each year, we included programs on the schedule Monday through Saturday from 8 p.m. to 11 p.m., and on Sunday from 7 p.m. to 11 p.m. Because some prime time program schedule changes or cancellations can occur in the fall prime time schedule, we used the debut schedule of programs that appeared in September of each selected year.

[57] Because data that track program production information were limited for programming on all cable networks, we looked at the ownership of basic cable networks only.

In: Media Industry Programming, Competition...   ISBN: 978-1-61122-078-0
Editors: Ryan E. Moore   ©2011 Nova Science Publishers, Inc.

Chapter 2

HOW THE SATELLITE TELEVISION EXTENSION AND LOCALISM ACT (STELA) UPDATES COPYRIGHT AND CARRIAGE RULES FOR THE RETRANSMISSION OF BROADCAST TELEVISION SIGNALS

Charles B. Goldfarb

SUMMARY

The Satellite Television Extension and Localism Act of 2010 (STELA), P.L. 111-175, modifies the copyright and carriage rules for satellite and cable retransmission of broadcast television signals. The legislation was needed to reauthorize (through December 31, 2014) certain expiring provisions in the Copyright Act and the Communications Act and to update the language in those acts to reflect the transition from analog to digital transmission of broadcast signals, as well as to address certain public policy issues. Had the expiring provisions not been reauthorized, satellite operators would have lost access to a statutory compulsory copyright license and to statutory relief from retransmission consent requirements. This would have made it difficult, if not impossible, for them to retransmit certain distant broadcast signals to their subscribers, including signals providing otherwise unavailable broadcast network programming.

The Copyright Act and Communications Act distinguish between the retransmission of local signals—the broadcast signals of stations located in the same local market as the subscriber—and distant signals. Statutory provisions block or restrict the retransmission of many distant broadcast signals in order to foster local programming. These provisions typically take the form of defining which households are "served" or "unserved" by local broadcasters, with unserved households eligible to receive distant signals. But there are many grandfather clauses and other exceptions built into the rules that allow households to receive otherwise proscribed distant signals. STELA generally retains, and in some cases expands upon, these grandfathered and exceptional cases.

STELA provides broadcasters two new incentives to use their digital technology to broadcast multiple video streams (to "multicast"). It clarifies that royalty fees are payable to copyright owners of the materials on non-primary digital voice streams as well as primary streams, thus encouraging broadcasters (who often hold some of those copyrights) to expand their multicasting. STELA specifically gives broadcasters the incentive to undertake such multicasting to offer otherwise unprovided network programming in so-called "short markets"—markets that do not have network affiliates for all four major networks. It does this by defining households that can receive the programming of a particular network from the non-primary multicast video stream of a local broadcaster as being served, rather than unserved, with respect to that network, thus prohibiting satellite operators from retransmitting to those households distant signals that carry that network's programming. The local broadcaster can then seek retransmission consent payments from satellite operators. Several other provisions in STELA also are intended to reduce the number of short markets or increase the flow of distant network signals into short markets.

Today, satellite operators are allowed, but not required, to offer subscribers the signals of the broadcast stations in their local market. The satellite operators have chosen not to offer this "local-into-local" service in many small markets, preferring to use their satellite capacity to provide additional high definition and other programming to larger, more lucrative markets. The costs associated with providing local-into-local service in small markets may exceed the revenues. STELA provides DISH Network, which currently is subject to a permanent court injunction that in effect prohibits it from retransmitting to its subscribers the signals of distant broadcast stations, the opportunity to have that injunction waived if it provides local-into-local service in all 210 local markets in the United States, which it began doing on June 3, 2010.

STELA does not address the issue of "orphan counties"—counties located in one state that are assigned to a local market, as defined by the Nielsen Media Research designated market areas, for which the principal city and most or all of the local broadcast stations are in another state.

OVERVIEW OF STELA

The Satellite Television Extension and Localism Act of 2010 (STELA), P.L. 11 1-175,[1] extends, updates, and modifies provisions in the Copyright Act[2] and the Communications Act[3] relating to the retransmission of broadcast television signals by satellite television and cable television providers. Among other things, STELA:

- Reauthorizes through December 31, 2014, expiring provisions that provide satellite carriers access to a simple statutory compulsory copyright license and free satellite carriers from retransmission consent requirements, when retransmitting to their subscribers the signals of certain broadcast stations located outside the subscribers' local markets ("distant signals"). Had these provisions expired, it would have been difficult, if not impossible, for satellite operators to provide to their subscribers broadcast network programming that the subscribers are unable to receive from their local broadcasters.
- Revises provisions in copyright and communications law to take into account the transition from analog to digital transmission of broadcast signals.
- Creates an incentive for broadcasters, who often hold copyrights on of the programming they broadcast, to use their digital capabilities to offer multiple video streams ("multicasting") by requiring satellite operators to pay royalty fees for the programming on the non-primary, as well as primary, video streams.
- Provides local broadcasters in markets that currently do not have network affiliates for all four major networks (so-called "short markets") the incentive to offer the programming of the currently unavailable networks on their non- primary digital video streams. Specifically, STELA defines households that can receive the

programming of a particular network from the non-primary multicast video streams of a local broadcaster as being "served" rather than "unserved" with respect to that network, thus prohibiting satellite operators from retransmitting to those households distant signals that carry that network's programming and allowing the broadcaster to seek retransmission consent payments.

- Frees DISH Network of a permanent court injunction against retransmitting the signals of distant network stations into short markets in exchange for the requirement to make available to its subscribers in *each of the 210 local markets* in the United States the signals of all the full-power broadcast stations in the local market. To meet that requirement, on June 3, 2010, DISH began providing such "local-into-local" service to the 29 local markets it had not been serving.

- Modifies the rules governing which households are eligible to receive distant signals from satellite carriers, generally grandfathering those households that currently receive such signals. These rule changes, which attempt to better reflectthe current market and technological environment, may increase the number of households that qualify to receive distant signals.

- Modifies the copyright administrative procedures, reporting requirements, royalty fees, filing fees, and non-compliance penalties for the improper retransmission of broadcast television signals by both satellite carriers and cable operators.

- Changes the statutory licenses applicable to the copyrighted material on the retransmitted signals of "significantly viewed" broadcast stations,[4] low power broadcast stations, and other statutorily exceptional[5] stations.

- Requires satellite operators to make available to their subscribers all the programming of non-commercial television stations that is in high-definition format.

- Requires the Register of Copyrights to submit a report on market-based alternatives to statutory licensing and also requires the Comptroller General to submit a report on changes to carriage requirements currently imposed on multichannel video programming distributors (MVPDs) and to Federal Communications Commission (FCC) regulations that might be required if Congress were to phase-out the current statutory satellite and cable licensing requirements.

STELA does not address the situation in which a county has been assigned to a local market for which the principal city is in another state and the television stations located in that local market primarily address the needs of households in that other state, rather than providing news, sports, and other programming of interest to the county. There have been a number of legislative proposals intended to address this "orphan county" issue, but none was included in STELA. But STELA does require the FCC to submit a report on the in-state broadcast programming available to households that receive the signals of broadcast stations that are considered, by statute and rule, to be local but are located in a different state.

BACKGROUND

Congress has constructed a regulatory framework for the retransmission of broadcast television signals by satellite television operators through a series of laws—the 1988 Satellite Home Viewer Act (SHVA),[6] the Satellite Home Viewer Act of 1994,[7] the 1999 Satellite Home Viewer Improvement Act (SHVIA),[8] the 2004 Satellite Home Viewer Extension and Reauthorization Act (SHVERA),[9] and most recently STELA. These laws have fostered satellite provision of MVPD service and, as satellite has become a viable competitor to cable television, have attempted to make the regulatory regimes for satellite and cable more similar. Today, the regulatory framework for satellite exists alongside an analogous, but in some significant ways different, regulatory framework for cable.[10]

The various provisions in these satellite acts created new sections or modified existing sections in the Copyright Act and the Communications Act of 1934. Under current law, in order to retransmit a broadcaster's signals to its subscribers, a satellite operator or a cable operator, with certain exceptions, must obtain a license from the copyright holders of the content contained in the broadcast for use of that *content* and also must obtain the consent of the broadcaster for retransmission of the broadcast *signal*. The statutory provisions addressing copyright are in the Copyright Act and are administered by the Copyright Office in the Library of Congress; those provisions addressing signal retransmission are in the Communications Act and are administered by the FCC. But in several cases, the provisions in one act are conditioned on meeting conditions prescribed in the other act or meeting rules adopted by the agency that administers the other act.

The satellite and cable regulatory frameworks attempt to balance a number of longstanding, but potentially conflicting, public policy goals—most notably, localism, competitive provision of video services, support for the creative process, and preservation of free over-the-air broadcast television. They also attempt to balance the interests of the satellite, cable, broadcast, and program content industries. Congress incorporated sunset provisions in SHVERA—and again in STELA—because of its concern that market changes could affect these balances. Indeed, as Congress debated the legislative proposals that were included in, or left out of, STELA, it gave substantial weight to a proposed package of changes in copyright procedures, royalty rates, and other parameters constructed and supported by a wide range of industry players through a process of direct negotiations and compromise.

The statutory provisions distinguish between the retransmission of *local* signals—the broadcast signals of stations located in the same local market (as defined by the 210 designated market areas into which the United States is divided by Nielsen Media Research) as the subscriber—and of *distant* signals. These provisions block or restrict the retransmission of many distant broadcast signals in order to protect local broadcasters from competition from distant signals and to provide them with a stronger negotiating position vis-à-vis the satellite and cable operators. The intent is to foster local programming. But the statutory framework also recognizes that U.S. households benefit from the receipt of certain distant broadcast signals and includes explicit retransmission and copyright rules for these.

The statutory framework for satellite sets the parameters within which industry players must conduct business. It provides answers to four fundamental business questions:

- May—or must—a satellite operator retransmit some or all local broadcast signals?[11]
- May a satellite operator retransmit certain categories of distant (non-local) broadcast signals?
- Is retransmission of those signals contingent on a satellite operator receiving the prior retransmission consent of—and providing compensation to—the broadcaster? and
- Is use of the content on those signals subject to specific copyright license terms?

Satellite operators and broadcasters also must conduct business within the constraints of longstanding industry practice. Broadcast program suppliers—

both broadcast networks and owners of non-network, syndicated programming—contractually grant individual broadcast television stations the exclusive broadcast rights to their programming in a geographic area and restrict those broadcast stations from allowing other parties to retransmit the station signals carrying that programming beyond the area of exclusivity. Thus, in some situations where the regulatory framework allows satellite (or cable) operators to retransmit the signals of a distant broadcast station, subject to obtaining the permission of the broadcast station, that station may be—and, in practice, often is—contractually prohibited from granting the MVPD retransmission consent.

Although satellite and cable operators compete directly with one another in most markets, there are significant differences in the regulatory frameworks under which they operate. These differences largely reflect the different origins of the cable and satellite industries—cable beginning as a business with technology focused on serving narrow geographic areas and satellite beginning as a business with technology serving broad geographic areas. To this day, cable network architecture and technology can more efficiently accommodate local programming than can satellite. Some observers have proposed that the retransmission, copyright, and other rules under which these competing multichannel video programming distributors operate should be rationalized to eliminate artificial competitive advantages or disadvantages. For example, the Copyright Office, in a report to Congress required by SHVERA ,[12] has proposed that the gross receipts royalty system for cable retransmission of distant broadcast signals in section 111 of the Copyright Act be replaced by a flat fee per subscriber system of the sort for satellite retransmission of distant broadcast signals in section 119 of the Copyright Act. The Copyright Office also has proposed[13] that the provisions defining satellite subscriber eligibility for receiving distant signals in section 119 (the "unserved household" provisions) be replaced by the imposition on satellite operators of the FCC's network non-duplication[14] and syndicated exclusivity rules,[15] which currently are used to limit the retransmission of distant broadcast signals by cable operators. But in the Congressional deliberations leading to passage of STELA, there was little discussion of a major modification of the regulatory framework.

Issues Addressed in STELA

Reauthorization

STELA extends through December 31, 2014, several statutory copyright and communications provisions, required for satellite operators to retransmit distant signals, that would have expired on May 31, 2010. Most significantly,

- Section 119 of the Copyright Act[16] provides satellite operators that retransmit certain "distant" (non-local) broadcast television signals to their subscribers with an efficient, relatively low cost way to license the copyrighted works contained in those broadcast signals—a statutory per subscriber, per signal, per month royalty fee. Had the law expired, it would have been very difficult (and perhaps impossible) for satellite operators to offer the programming of broadcast networks[17] to that subset of subscribers who currently cannot receive that programming from local broadcast stations that are affiliated with those networks.[18] It also would have been difficult for satellite operators to offer their subscribers the signals of distant stations that are not affiliated with broadcast networks, including both "superstations"[19] and other non-network stations.

- In addition, prior to the enactment of STELA, section 119 provided those satellite operators that retransmit to their subscribers the signals of "significantly viewed" stations—stations that are located outside the local market in which the subscriber is located but have been determined to be "significantly viewed" by those households in the local market that do not subscribe to any MVPD provider—a royalty-free license for the copyrighted works contained in those broadcast signals. Had section 119 expired, it would have been very difficult (and perhaps impossible) for satellite operators to offer their subscribers the signals of significantly viewed stations. Under STELA, satellite retransmission of significantly viewed stations has been moved from section 119 to section 122 of the Copyright Act, under which such retransmission is subject to the royalty-free license in section 122.

- Section 325(b)(2)(C) of the Communications Act[20] allows a satellite operator to retransmit the signals of distant network stations, without first obtaining the retransmission consent of those distant stations, to those subscribing households that cannot receive the signals of local broadcast television network affiliates. Had it expired, a satellite operator would have had to negotiate compensation terms with those distant network stations whose signals it retransmitted to those "unserved" subscribers.

- Section 325(b)(3)(C)(ii) of the Communications Act[21] prohibits a television broadcast station that provides retransmission consent from engaging in exclusive contracts for carriage or failing to negotiate in good faith. Section 325(b)(3)(iii)[22] prohibits an MVPD from failing to negotiate in good faith for retransmission consent. Had these provisions expired, a broadcaster or an MVPD could have chosen to employ a "take it or leave it" strategy rather than to negotiate retransmission consent terms in good faith, increasing the risk of an impasse that results in subscribers losing access to the broadcast station's programming.

STELA includes a provision making the effective date of the act February 27, 2010, in order to protect satellite operators from potential lawsuits for copyright infringement for the brief period of time when the old authorization had expired and Congress had not yet enacted new authorization. At that time, Congress had encouraged the satellite operators not to discontinue retransmission of the distant signals in order to allow satellite subscribers to continue to receive those signals.

Revising Existing Rules That Are Based on Analog Technology

A number of statutory provisions, and many FCC and Copyright Office rules adopted to implement statutory provisions, have been based on the transmission of analog broadcast signals, but during 2009 the transition to digital broadcast signals was largely achieved. As a result, statutes and rules that explicitly referred to analog technology were no longer effective in attaining the objectives for which they were enacted. Thus, Marybeth Peters, Register of Copyrights, proposed five modifications to section 111 of the Copyright Act and four modifications to section 119 of the Copyright Act "to accommodate the conversion from analog to digital broadcasting."[23] Analogous changes were proposed for the Communications Act.

STELA includes specific changes to language in the Copyright Act and to the Communications Act intended to make them consistent with a digital environment. It also includes provisions directing the FCC to develop a predictive model for the reception of digital signals within six months of enactment in order to determine which households are "unserved" and therefore eligible to receive digital network signals. STELA also includes a provision that provides guidance for the period before the new predictive model has been implemented.

STELA modifies the methodology used to determine whether a household is served to reflect the current market and technological environment, including the transition from analog to digital transmission. It is possible that some of the methodological changes may increase the number of households eligible to receive distant network signals.[24] For example, most households now receive their broadcast signals from their cable or satellite service rather than over-the-air and therefore do not use a rooftop antenna. The old definition of unserved household referred to the inability to receive a signal of a specified intensity using a rooftop antenna; STELA changes the definition to refer to *any* antenna. Since indoor antennas, such as "rabbit-ear" antennas, tend to be less effective than rooftop antennas, this may increase the number of households that qualify as unserved.

Fostering Digital Multicasting, Especially Multicasting to Provide Network Programming in Those Markets That Lack a Network Affiliate ("Short Markets")

Although each of the four major broadcast television networks (ABC, CBS, FOX, and NBC) has a local station affiliate in most U.S. markets, 58 of the 210 markets do not have the full complement of four network affiliates.[25] In these short markets, subscribers have been defined as being "unserved" with respect to the missing network and satellite operators have been allowed to retransmit to their subscribers the signals of up to two distant stations that are affiliated with that missing network.[26]

With the transition from analog to digital technology, however, broadcast stations are able to broadcast multiple video streams. Some local television stations in short markets are affiliated with a national network and broadcast that network's programming on their primary video stream, but also have reached agreements with a second national network that lacks an affiliate in the local market to carry the network programming of that second network on a non-primary video stream. This multicasting allows households in the local market to receive the network programming of that second network, although it is unlikely that the local station provides any original local programming on that secondary video stream.

Under STELA, if a local television station broadcasts a non-primary video stream that provides the programming of a national network and was carried by a satellite operator on March 31, 2010, and if the local station continues to carry that network's programming on that video stream, then as of October 1, 2010, that video stream will be considered a "qualified multicast video" and households in that local market will be considered served with respect to the broadcast network whose programming is carried on that video stream. Thus, after October 1, 2010, a satellite operator could not use the statutory distant signal copyright license to retransmit to households in that local market the signal of a distant broadcast station affiliated with that streamed network. Presumably, the satellite operator would have to obtain retransmission consent from the local broadcaster (which probably would entail making a payment to the broadcaster) to retransmit the programming as part of its local-into-local service.

As of January 1, 2011, all non-primary video streams of national network programming offered by a local television station will be considered qualified multicast video and households in the local market will be considered served with respect to the broadcast network whose programming was carried on those video streams.[27] As a result of this change in treatment of network programming broadcast over non-primary video streams, satellite operators would be allowed to retransmit the programming as part of their local-into-local service offering (if they successfully negotiated a retransmission consent agreement with the broadcaster), but would no longer be able to retransmit that network programming using a distant broadcast signal.

STELA allows a satellite subscriber who was lawfully receiving the distant signal of a network station on the day before enactment of the new legislation to receive both that distant signal and the local signal of a network station affiliated with the same network until the subscriber chooses to no longer receive the distant signal from its satellite operator. Thus, if in a short market a local broadcaster began to multicast on a non-primary video stream the programming of the network for which there has been no local affiliate, and the satellite operator chose to retransmit that non-primary video stream, a subscriber who has been receiving the distant network signal could continue to receive that distant signal as well as the local network signal, as long as the subscriber did not discontinue its subscription for that distant signal. A household in that short market would not be allowed to receive a distant network signal, however, if it received from the satellite operator the programming of that same network from the non-primary video stream of a local broadcaster but was not a subscriber lawfully receiving the distant signal on the day before enactment of the new legislation.

Another provision in STELA fosters multicasting in all markets, not just short markets. It encourages broadcasters to offer programming over multiple digital video streams—both their primary stream and non-primary streams— by clarifying that satellite operators must pay copyright royalty fees for the retransmission of the programming on broadcasters' non-primary as well as primary video streams. Since broadcasters often hold some copyrights for the programming they broadcast, such payments increase their incentive to multicast.

Providing an Incentive for DISH Network to Offer Local-into-Local Service in All Designated Market Areas: Allowing DISH to Use a Statutory License to Retransmit Distant Network Signals into Short Markets

Satellite operators are allowed, but not required, to offer subscribers the signals of all the broadcast stations in their local market. If a satellite operator chooses to retransmit the signal of a local broadcast station and to take advantage of a royalty-free statutory copyright license for the content carried on that signal, it must retransmit the primary signals of all the full power stations in that local market, subject to obtaining local station permission. The satellite operators have chosen not to offer this "local-into-local" service in many small markets, preferring to use their satellite capacity to provide additional high definition and other programming to larger, more lucrative markets than to use the capacity to serve very small numbers of customers. In some cases, those small markets may not generate enough revenues to cover the costs of providing local-into-local service.[28] As a result, approximately 3% of all U.S. households do not have access to any local broadcast signals if they subscribe to satellite video service, unless they can receive those signals directly over-the-air.[29]

Early in the 111th Congress, Representative Stupak introduced H.R. 927, the Satellite Consumers' Right to Local Channels Act, which would, in effect, require satellite operators to offer local-intolocal service in all markets; if a satellite operator wished to use the royalty-free statutory copyright license to rebroadcast the content on a broadcast signal in *any* local market, it would have to provide local-into-local service in *every* market. But during markup of the House Energy and Commerce Committee bill, H.R. 2994, Representative Stupak agreed to withdraw his bill (which he had introduced in the form of an amendment), when DISH Network indicated that it would voluntarily provide local-into-local service in all 210 markets within two years in exchange for statutory relief from a current court injunction prohibiting it from providing its subscribers distant signals using the section 119 copyright license.[30] That *quid pro quo* has been incorporated into STELA.

As a result of repeated violations of section 119 of the Copyright Act, DISH Network currently is subject to a permanent injunction, imposed by the U.S. Court of Appeals for the 11th Circuit,[31] barring it from using the section 119 statutory license for the copyrighted materials when retransmitting distant signals to its subscribers; it therefore must employ an arms-length agreement with National Programming Service for that entity to deliver distant signals to its subscribers. Under STELA, the injunction would be partially waived if DISH Network provides local-into-local service in all 210 local markets in the United States. Specifically, DISH would be allowed to use a section 119 license for the copyrighted materials when retransmitting to its subscribers in a "short market" the signals of a distant network broadcast station affiliated with a network for which no local broadcaster is providing the network programming over its primary video stream.

Because of DISH's long history of illegally retransmitting distant signals, STELA incorporates a number of safeguards. DISH must demonstrate that it is offering local-into-local service in all 210 local markets in the United States (referred to as designated market areas or DMAs) in order to be deemed qualified by the court for a temporary waiver of the injunction. The Court must select a special master who would make an initial examination and provide on-going monitoring to assure that DISH is serving all 210 DMAs (and if not, make a determination that it is nonetheless acting reasonably and in good faith) and is in compliance with the royalty payment and household eligibility requirements of the license. The initial waiver of the injunction would be temporary, but could be extended for good cause; if DISH lost recognition as a qualified carrier it could not seek to be re-qualified. Also, the Comptroller General is instructed to monitor the degree to which DISH is complying with the special master's examination. DISH would have the burden of proof that it is providing local-into-local service with a good quality satellite signal to at least 90% of the households in each DMA. It would be subject to penalties of between $250,000 and $5 million for failure to provide service, with exceptions for nonwillful violations.

On June 3, 2010, DISH introduced local-into-local service in the 29 DMAs it had not been serving. These markets are: Alpena, MI; Biloxi, MS; Binghamton, NY; Bluefield, WV; Bowling Green, KY; Columbus, GA; Elmira, NY; Eureka, CA; Glendive, MT; Greenwood, MS; Harrisonburg, VA; Hattiesburg, MS; Jackson, TN; Jonesboro, AR; Lafayette, IN; Lake Charles, LA; Mankato, MN; North Platte, NE; Ottumwa, IA; Parkersburg, WV; Presque Isle, ME; Salisbury, MD; Springfield, MA; St. Joseph, MO; Utica, NY; Victoria, TX; Watertown, NY; Wheeling, WV, and Zanesville, OH.

STELA also requires each satellite carrier to submit a semi-annual report to the FCC setting forth (1) each market in which it offers local-into-local service; (2) detailed information regarding the use of satellite capacity for the provision of local-into-local service; (3) each local market in which it has commenced offering local-into-local service in the six-month period covered by the report; and (4) each local market in which it has ceased to offer local-into-local service in the six- month period. The FCC is required to submit to the Senate Commerce, Science, and Transportation Committee and the House Energy and Commerce Committee within a year a report containing the findings, conclusions, and recommendations of a study of (1) incentives that would induce a satellite carrier to offer local-into-local service in markets the carrier has not been serving; and (2) the economics and satellite capacity conditions affecting delivery of local signals by satellite carriers to these markets.

Reducing the Number of Short Markets by Eliminating the "Grade B Bleed" Problem

Prior to enactment of STELA, in areas where a network-affiliated broadcast station was located near the DMA boundary, so that its signal extended into a portion of a neighboring DMA that did not have a local station affiliated with the same network, households in that neighboring market who could receive that signal at a Grade B level were not considered to be "unserved" for that network. A satellite operator could neither offer that overlapping signal to those households as part of local-into-local service (since it was a distant signal) nor provide to those households the signal of a distant station affiliated with the same network, because those households were not considered unserved. The satellite operators sought to eliminate this so-called "Grade B bleed" problem by modifying the test for a subscriber being unserved to apply only to the strength of the signal from an in-market station or by defining unserved in terms of whether the viewer can get local service from the satellite spot beam, rather than in terms of over-the-air reception.[32]

STELA eliminates the problem by defining as "unserved" those households that do not receive the network programming from an over-the-air signal that originates in the local market, that is the signal of their *local network affiliate.*

Household Eligibility to Receive Distant Signals:Grandfathered Subscribers, Other Subscribers, and Households That Are Not Subscribers When Legislation Is Enacted ("Future Applicability")

The primary mechanism for limiting satellite retransmission of distant network signals has been to restrict such retransmission to "unserved" households that cannot receive the programming of a particular network because either (1) the satellite operator is not offering local-into-local service in that market and the households cannot receive a signal of a threshold quality level over-the-air from the local network affiliate, or (2) there is no local affiliate offering the programming of that network. But both the Copyright Act and the Communications Act include certain grandfathered exceptions to those eligibility restrictions; as a result, many households that are able to receive a network signal from a local broadcast station are allowed to continue to receive the distant signal of a broadcast station affiliated with the same network. STELA retains most of these grandfathered exceptions and in some ways expands on them.

Section 339 of the Communications Act sets the rules for carriage of distant television station signals by satellite operators. Section 339(a)(2) addresses the replacement of distant signals with local signals, enumerating four different sets of rules: for grandfathered subscribers to analog distant signals, for other subscribers to analog distant signals, for households that are not subscribers at the time the legislation is enacted (future applicability), and for subscribers to distant digital signals. STELA:

- Retains section 339(a)(2)(A), the grandfathering provision that allows certain households that historically had been receiving distant network signals illegally (and therefore otherwise would not have qualified to receive those distant signals) to continue to receive those signals. The language has been updated only to reflect the date of enactment of the new legislation and to eliminate reference to analog technology. All these households continue to be grandfathered to receive distant network signals despite being able to receive the signals of local stations with the same network affiliation.

- Eliminates references to analog signals from section 339(a)(2)(B), but otherwise the two-part provision is retained. Under the first part, if a household's satellite operator had made a local network station available on January 1, 2005, as part of local-into-local service, the operator would nonetheless be allowed to provide to that household a distant signal of a station affiliated with the same network if the operator had submitted to the television network no later than March 1, 2005, a list of households receiving that distant signal that included that household. This continues the grandfathering of households that had been legally receiving a distant network signal and were allowed to continue to receive that signal when they also had access to the signal of a local broadcast station affiliated with the same network. Under the second part, if the satellite operator had not made available a local network station on January 1, 2005, as part of local-into-local service, the operator would be allowed to offer the household the distant network signal only if (a) the household seeks to subscribe to the distant signal before the date on which the operator begins to offer local-into-local service, and (b) the operator submits to each television network within 60 days of commencing such service the households subscribing to the distant signal. Thus, a household that had legally sought to receive a distant network signal is allowed to continue to receive that signal after the signal of a local broadcast station affiliated with the same network is available.

- Allows a subscriber who is lawfully receiving the distant signal of a network station from a satellite operator on the day before enactment of STELA to receive both the distant signal and the local signal of the same network until the subscriber chooses to no longer receive the distant signal from the satellite operator (whether or not the subscriber elects to subscribe to local-into-local service). Thus, all the households legally receiving distant network signals under section 339(a)(2)(B) at the date of enactment of STELA continue to be allowed to receive those distant signals.

- Prohibits a satellite operator from providing a distant network signal to a person who (1) (a) is not a subscriber legally receiving that distant signal on the date STELA is enacted, and (b) at the time the person seeks to receive the distant signal, resides in a local market where the satellite operator offers local-into-local service that includes a local station affiliated to the same network and the person can receive that local-into-local service, *or* (2) (a) is a subscriber legally receiving a distant signal on or after the date STELA is enacted, and (b) subsequent to such subscription the satellite carrier makes available to that subscriber the signal of a local network station affiliated with the same network as the distant signal (and the retransmission of such signal by the carrier can reach the subscriber), unless the person subscribes to the signal of the local network station within 60 days after the signal is made available. The latter is intended to support local stations by requiring the subscriber to obtain local-intolocal service in order to continue to receive the distant network signal.

- Defines a subscriber as eligible to receive a distant signal of a network station affiliated with the same network as a local station if, with respect to a local network station: (1) the subscriber's household is not predicted by the model specified in the act to receive the threshold signal intensity; (2) the household is determined, based on a test conducted in accordance with the current model or any successor regulation, not to be able receive the signal of the local station with an intensity that exceeds the standard; or (3) the subscriber is in an unserved household as determined by the definition of an unserved household in section 11 9(d)(1 0)(A) of the Copyright Act. The third criterion appears to allow a household that does not meet the signal intensity test for analog service for the signal of a local network station to be grandfathered for the receipt of a distant network signal carrying the same network, even if that household could receive the digital signal of the local network station.

Provisions in section 119 of the Copyright Act define "unserved households" and set the copyright rules that apply to the secondary transmission of distant signals to those unserved households. STELA modifies some of those provisions.

- If a local station is multicasting and offers a second network's programming on one of its non-primary video streams, but a household using an antenna cannot receive that non-primary video stream at the signal intensity specified in FCC rules, then the household is deemed unserved with respect to the network whose programming is being broadcast on that non-primary stream. This takes effect on October 1, 2010, for multicast streams that existed on March 31, 2010, and on January 1, 2011, for all other multicast streams.

- References to analog signals are eliminated, but otherwise all the rules covering grandfathered households receiving distant signals currently in section 11 9(a)(4)(A) are retained.

- For a subscriber (other than a grandfathered household) who, on the day before enactment of STELA, was lawfully receiving a satellite retransmission of a distant network signal under a statutory license, the statutory license shall apply for the retransmission of that distant signal. Further, the subscriber's household shall continue to be considered an unserved household with respect to that network until the subscriber elects to stop receiving that distant signal, whether or not the subscriber has access to the signal of a local network station affiliated with the same network through local-into-local service and whether or not the subscriber elects to subscribe to that local-into-local service. This, in effect, creates a new group of grandfathered households.

- The statutory distant signal copyright license in section 119 of the Copyright Act does not apply to the satellite retransmission of a distant network signal to a person who is not a subscriber lawfully receiving that distant network signal at the date of enactment of STELA if, when that person subsequently seeks to subscribe to a satellite carrier for that distant signal, that person can obtain that network's programming from a local station affiliated with the same network through local-into-local service.

- The statutory distant signal copyright license in section 119 of the Copyright Act applies to the satellite retransmission of a distant network signal to a person who is a subscriber lawfully receiving that distant network signal on or after the date of enactment of STELA, and the subscriber's household continues to be considered to be an unserved household with respect to that network, until such time as the subscriber elects to terminate such retransmission, but only if the person subscribes to retransmission of a local network station affiliated with the same network (that is, subscribes to local-into-local service) within 60 days of the satellite carrier making local-into-local service available to the subscriber. Thus, a household can be grandfathered for the distant network service only if it subscribes to local-into-local service within 60 days of that service becoming available.

Modified Copyright Treatment of the Satellite Retransmission of Low Power Television Station Signals

Low power television service was created by the FCC in the 1980s to serve small communities (rural or urban) with low cost, limited geographic range facilities that used available spectrum between full power stations. It is a "secondary service" that is not guaranteed protection from interference or displacement by full service stations. Low power stations that produced at least two hours per week of local programming, maintained a production studio within their Grade B contour, and complied with many of the requirements placed on full service stations were given a one-time opportunity to obtain "Class A" status that gave them primary status, that is, protected their channel from interference or displacement.

Historically, satellite retransmission of low power television signals was covered by the statutory distant signal copyright license in section 119 of the Copyright Act. Satellite operators were allowed to retransmit the signals of low power stations to subscribers within certain geographic limitations—to subscribers within 20 miles of the station transmitter for network-affiliated stations located in the 50 largest markets, within 35 miles of the station transmitter for network- affiliated stations located in other markets, and within the same designated market area as nonnetwork-affiliated stations.[33] Satellite operators had no copyright royalty obligation for retransmission of the low power station content within those same mileage limits; beyond those limits, satellite operators were subject to the statutory copyright license fees for distant signals outlined in section 119 of the Copyright Act.[34]

Under STELA, if satellite operators seek to use a statutory license for the copyrighted material on the low power television stations whose signals they retransmit, they must use the royalty-free statutory local signal license in section 122, rather than the section 119 license. STELA expands the geographic area covered by the royalty-free statutory license to the entire DMA in which the low power station is located.

STELA also explicitly states that a satellite carrier that retransmits the signal of a low power station under a statutory license is not required to make any other secondary retransmissions. Thus retransmission of a low power station does not trigger the requirement to offer local-into-local service or to retransmit any other low power stations. No local low power station can demand carriage by the satellite operator serving its market area, even if that satellite operator is providing local-into-local service.

Since low power television stations do not have a deadline for their transition from analog to digital transmission, the old, analog-based FCC rules for determining whether a household is eligible to receive distant signals will apply with respect to low power television until the station is licensed to broadcast a digital signal.

The statutory local signal copyright license does not apply to satellite retransmission of repeaters or translators.

Satellite Carriage of Noncommercial Educational Television Stations

By statute, providers of direct broadcast satellite service (DirecTV and DISH Network) must reserve between 4% and 7% of their channel capacity exclusively for noncommercial programming of an educational or informational nature.[35] With the digital transition, broadcasters now are able to broadcast high definition signals and multiple digital programming streams over their licensed spectrum, and the public television stations are seeking to expand satellite carriage of their high definition and multicast signals.

The public broadcasters have reached retransmission consent agreements with DirecTV, the cable industry (through both the National Cable and Telecommunications Association representing large cable operators and the American Cable Association representing small cable operators), and Verizon for the retransmission of most of their high definition and multicast video streams. The agreement with DirecTV incorporated "creative solutions that recognized [DirecTV's] capacity limitations; ultimately ensuring that subscribers have access to the myriad of content and services provided by the local stations while accommodating their capacity concerns."[36] The public broadcasters have not yet achieved retransmission agreement with DISH Network, but negotiations are continuing.

STELA modifies section 33 8(a) of the Communications Act, which addresses the carriage of local television signals by satellite carriers, to require any satellite carrier that has not yet negotiated a carriage contract covering at least 30 noncommercial educational television stations—that is, DISH Network—(1) to provide subscribers, by the end of 2011, the high definition signals of qualified noncommercial educational television stations in all the local markets in which the carrier currently offers local television broadcasts in high definition (and by the end of 2010 to half of those markets), and (2) when initiating the provision of high definition local broadcast television in a market, to include the high definition signals of all qualified local noncommercial educational television stations.

At an October 7, 2009, hearing of the Senate Subcommittee on Communications, Technology, and the Internet, public broadcasters identified another problem for which they sought a legislative solution. Most states have developed state public television networks intended to serve the entire state, but in 16 states those networks do not have public stations transmitting signals in each DMA in the state; under current law, satellite carriers are not allowed to use a royalty-free statutory copyright license to retransmit the signals of the in-state, but out-of-market public broadcasting stations to their subscribers in those DMAs.[37] STELA modifies the provisions for the royalty-free statutory copyright license in section 122 of the Communications Act to allow, where there is a public educational network of three or more noncommercial educational broadcast stations in a state, a satellite operator to use the royalty-free license to retransmit the programming on those stations' signals to subscribers in any county in the state whose households are otherwise ineligible to receive retransmissions of those signals.

Satellite Carriage of State Public Affairs Networks

Cable franchise authorities are allowed, by law, to require cable operators to set aside some of their capacity for the carriage of public, educational, and governmental (PEG) programming. This programming is not broadcast to the public; it is sent directly to the cable system's head-end for retransmission. Satellite operators are not required to offer PEG programming, though they have the obligation to allocate between 4 and 7 percent of their channel capacity exclusively to noncommercial programming of an educational or informational nature. In order to foster PEG programming on cable systems, a number of states have created state public affairs networks that produce non-broadcast programming of state-wide interest. Although this programming is available to satellite operators, those operators are not widely offering it to subscribers.

STELA includes a provision intended to encourage satellite operators to carry these state public affairs networks. Under the provision, a satellite carrier that provides the retransmission of the state public affairs networks of at least 15 different states, under reasonable prices, terms, and conditions, and does not delete any of the noncommercial educational or informational programming on those networks, would only have to allocate 3.5% of its channel capacity to the retransmission of educational or informational noncommercial programming, rather than 4%. This provision might encourage satellite operators to offer state public affairs networks to subscribers in orphan counties or in short markets.

The Retransmission of In-State, but Non-Local, Broadcast Signals into Counties Assigned to Local Markets in Other States ("Orphan Counties")

The current regulatory frameworks for both satellite and cable distinguish between the retransmission of local and distant signals and require that local markets be defined by the DMAs constructed and published by Nielsen Media Research. [38] The viewing patterns that underlie these Nielsen markets are primarily the result of the physical locations of the various broadcast television stations and the reach of their signals. (They also reflect the boundaries of the exclusive broadcast territories that each of the three original television broadcast networks—ABC, CBS, and NBC—had incorporated into their contracts with their local affiliate stations decades ago.) DMAs do not take into account state boundaries. As a result, under current statutes and rules, a number of counties are assigned to local markets for which the principal city (from which all or most of the local television signals originate) is outside their state.[39] Satellite subscribers (and many cable subscribers) in these "orphan counties" may not be receiving signals from in-state broadcast stations and as a result may not be receiving news, sports, and public affairs programming of interest in their state.

Many residents of orphan counties have proposed that the statutory framework be modified to remove prohibitions or impediments on satellite operators retransmitting to their subscribers in these counties the signals of broadcast stations in in-state, but non-local, markets. (SHVERA selectively removed these impediments through four "exceptions" that allow satellite operators to retransmit to their subscribers in particular orphan counties in New Hampshire, Vermont, Oregon, and Mississippi—but not in other locations—the signals of in-state but out-of-market broadcast stations.[40]) Broadcasters, however, have voiced concern that allowing such retransmission could undermine their financial viability by reducing their audience share and thus reducing their advertising revenues. They also assert such retransmission would weaken the local broadcasters' negotiating position with the satellite and cable operators, who could turn to the programming of an in-state but out-of-market affiliate of a particular network if they failed to reach retransmission consent with the local affiliate of that network. Broadcasters claim this would harm their ability to provide quality local programming, which is expensive to produce.[41]

A number of bills have been introduced in the 111th Congress that directly address this issue (either generically or for specific states or geographic areas) by allowing satellite operators to retransmit to subscribers in orphan counties the signals of certain in-state, but non-local broadcast stations.[42] But STELA (reflecting each of the four bills that had been reported out of the House Energy and Commerce, House Judiciary, Senate Commerce, Science, and Transportation, and Senate Judiciary committees, leading to STELA) does not include any provisions that would address this issue directly. During the markup of the Senate Judiciary Committee bill (S. 1670), reportedly several Senators prepared amendments that would have narrowly addressed the orphan county issue in their states, but then agreed to withdraw their amendments when other Senators voiced concern that the provisions would delay passage of the legislation because of unresolved issues among broadcasters and satellite operators. At the markup, reportedly there was discussion of imposing a deadline on the industry to reach a negotiated solution, such as a proposal by Senator Coburn that, if there were no industry agreement by the time the legislation reaches the Senate floor, a trigger provision would be inserted in the bill that would impose a statutory solution for the orphan counties issue if no negotiated compromise is reached after two years.[43] But STELA does not include a trigger provision.

STELA includes a provision instructing the FCC to prepare within one year a report containing analysis of (1) the number of households in each state that receive local broadcast signals from stations of license located in a different state; (2) the extent to which consumers have access to instate broadcast programming; and (3) whether there are alternatives to the use of DMAs to define local markets that would provide more consumers with in-state broadcast programming.

In addition, a savings clause in STELA—stating that nothing in the legislation, in the Communications Act, or in any FCC regulation shall limit the ability of a satellite operator to retransmit a performance or display of a work pursuant to an authorization granted by the copyright owner—is intended to clarify that a satellite operator always has the opportunity to negotiate a copyright license outside the section 119 statutory license. This clarification is not likely to result in the satellite retransmission into orphan counties of the sports and network programming on in-state, but out-of-market stations, but could encourage the retransmission of those stations' locally produced news programming.

Changing the Statutory Copyright License Applied to the Content on the Signals of Significantly Viewed and "Exception" Broadcast Stations

The statutory framework for the retransmission of broadcast television signals has been based on a distinction between local and distant signals. The signals of significantly viewed stations and the signals of in-state, out-of-market stations in the four states that satellite operators were allowed to import into orphan counties under the exceptions in SHVERA, originate outside the market into which they are imported; in that regard, they are distant signals and they have been subject to the section 119 distant signal statutory copyright license. But since significantly viewed stations and the "exception" stations can be presumed to be providing programming of local or state-wide interest to counties in particular local markets, arguably that content could be viewed as local to the counties into which they are imported and should be treated accordingly. STELA modifies the Copyright Act to treat those signals as local, moving the relevant provisions from section 119 to section 122.

STELA changes language in the heading of section 122 from "secondary transmission by satellite carriers within local markets" to "secondary transmission of local television programming by satellite." It makes satellite retransmission of both significantly viewed stations and the exception stations subject to the local signal statutory copyright license in section 122 rather than the distant signal statutory license in section 119, but requires the satellite operator to continue to pay the statutory copyright license fees under section 119 for the retransmission of the exception stations. Since significantly viewed stations already are subject to the royalty-free license in section 122, effectively there is no change in copyright treatment for the content on the signals of significantly viewed stations. But the statutory change allows DISH Network, which currently is under a court injunction prohibiting it from using the section 119 statutory copyright license to retransmit the content of broadcast signals, to use the section 122 statutory copyright license to do so.

Although STELA changes the statutory license required for satellite retransmission of the signals of significantly viewed and exception stations, it does not affect the retransmission consent requirement or the exemption from the FCC's network non-duplication and syndicated exclusivity rules, as they currently apply to significantly viewed and exception stations. It does, however, include a provision stating that the satellite operator would not be required to carry the significantly viewed stations or exception stations if it offered local-into-local service.

Allowable Signal Formats for the Retransmission of Significantly Viewed Stations

The satellite operators have complained that although both cable and satellite operators may offer significantly viewed stations, only satellite operators have been subject to an "equivalent bandwidth" provision that, as interpreted by the FCC, required the satellite operator to carry the signals of a significantly viewed station that is affiliated to the same network as a local station in the same format as that local station every moment of the day. Thus, for example, if the local station were not transmitting its programming in high definition format, the satellite operator would not be allowed to retransmit into the market the signals of the significantly viewed station in high definition format. According to satellite operators, this was infeasible.

STELA clarifies that a significantly viewed signal may only be provided in high definition format if the satellite carrier is passing through all of the high definition programming of the corresponding local station in high definition format as well; if the local station is not providing programming in high definition format, then the satellite operator is not restricted from providing the significantly viewed station's signal in high definition format.

Studying What the Impact Would Be If the Statutory Licensing System for Satellite and Cable Retransmission of Distant Broadcast Signals Were Eliminated

The United States Copyright Office has proposed that Congress abolish sections 111 and 119 of the Copyright Law, arguing that the statutory licensing systems created by these provisions result in lower payments to copyright holders than would be made if compensation were left to market negotiations.[44] According to the Copyright Office, the cable and satellite industries no longer are nascent entities in need of government subsidies, have substantial market power, and are able to negotiate private agreements with copyright owners for programming carried on distant broadcast signals.

One possible way to transition from the current licensing system would be to enact a statutory "trigger" mechanism, by which once a broadcast station successfully demonstrated that it had obtained the rights to negotiate for all the holders of copyrighted materials on its programming, so that a satellite carrier did not have to negotiate with multiple copyright holders, the statutory license for that station would sunset and the satellite operator would have to undertake private negotiations. This is strongly opposed by satellite operators, who question how voluntary licensing arrangements and sublicensing would work in practice.[45] Other parties argue that the current licensing systems are efficient and that the purpose of copyright law is to balance the potentially conflicting goals of fostering the dissemination of copyrighted material and allowing the copyright holder to be compensated by giving the copyright holder a *limited* monopoly over its material; they oppose a rule that allows the copyright holder to fully exploit its monopoly power to receive whatever the market would bear.[46]

STELA instructs the Copyright Office, after consultation with the FCC, to submit to the House and Senate Judiciary Committees, within one year, a report containing proposed mechanisms, methods, and recommendations on how to implement a phase-out of the current statutory license requirements in sections 111, 119, and 122 of the Copyright Act, including recommendations for legislative or administrative actions.

STELA also instructs the Comptroller General to prepare and submit a report within 12 months that analyzes and evaluates the changes to the cable and satellite carriage requirements in the Communications Act and in FCC rules that would be required if Congress implemented a phaseout of the current section 111, 119, and 122 statutory licensing requirements in the Copyright Act. It instructs the Comptroller General to consider the impact of such a phase-out on consumer prices and access to programming and to include recommendations for legislative or administrative actions.

Providing Digital Service on a Single Dish

Under section 338(g) of the Communications Act, satellite operators have been required to provide to their subscribers the analog signals of all broadcast stations on a single roof-top dish. Operators have been allowed to use a second dish for the provision of digital signals, but there was no requirement that all digital signals be provided on the same dish. STELA modifies section 33 8(g) to require a satellite operator, if it offers local-into-local service in a market, to provide to a subscriber the digital signals of all the local broadcast stations on a single dish.

Modification of the Methodology for Setting Copyright Royalty Rates and of Copyright Administrative Procedures and Requirements

STELA modifies the methodology for setting copyright royalty rates as well as copyright administrative procedures and requirements. Among these changes, STELA:

- requires satellite operators whose retransmissions of distant broadcast signals are subject to the section 119 statutory license to pay a filing fee, to be determined by the Register of Copyrights, to help recoup the administrative costs of distributing royalty fees;
- modifies the section 119 statutory royalty fee payable to copyright owners to take into account the non-primary streams of multicasting broadcasters;
- instructs the Register of Copyrights to issue regulations to permit interested parties to verify and audit the statements of account and royalty fees submitted by satellite carriers and cable operators;
- changes the process for adjusting royalty fees. Most significantly, STELA creates a proceeding of the Copyright Royalty Judges, which replaces the previously used compulsory arbitration proceeding, to determine royalty rates;

- instructs the Copyright Royalty Judges, when determining royalty rates in those situations where the parties are not able to reach a negotiated agreement, to establish fees that represent the fair market value of the retransmissions, basing their decision on economic, competitive, and programming information presented by the parties;
- requires the Copyright Royalty Judges to make an annual adjustment to the royalty fee based on the most recent consumer price index for all consumers and for all items;
- increases the statutory maximum damages to be imposed on satellite operators for violating territorial restrictions on the retransmission of distant broadcast signals from $5 to $250 per subscriber per month during which the violation occurred. It also increases the maximum statutory damages for regional or large- scale violations (that do not trigger a permanent injunction) from $250,000 for each 6-month period to $2.5 million for each 3-month period. One half of the statutory damages ordered are to be deposited with the Register of Copyrights and distributed to copyright owners;
- modifies the statutory license for retransmission by cable systems in section 111 of the Copyright Act by increasing the specified percentages of the gross subscriber receipts that cable operators must pay;
- updates the definition of "distant signal equivalent" used to reflect and take into account multicast signals when calculating the cable royalty payment, and sets a schedule for when these changes go into effect based on existing contractual agreements;
- clarifies that the royalty rates specified in sections 256.2(c) and (d) of title 37, Code of Federal Regulations, commonly referred to as the "3.65% rate" and the "syndicated exclusivity surcharge," respectively, do not apply to multicast streams;
- clarifies that when a cable operator retransmits a distant broadcast signal to a service area comprised of multiple communities, in which some communities are permitted to receive that signal and other communities are prohibited to do so, the royalty calculation does not include payment for the households that are not allowed to receive the signal;[47]
- modifies the methodology for determining the maximum and minimum royalty payments for small cable systems; and
- creates filing fees for satellite carriers and cable operators filing statements of account for section 111, 119, and 122 statutory

copyright licenses that are reasonable and that do not exceed one-half of the cost incurred by the Copyright Office for the collection and administration of the statements of account and any royalty fees deposited with the statements.

Severability

STELA includes a "severability" provision stating that if any provision of the new law, amendment made by the new law, or applications of such provision or amendment is held to be unconstitutional, the remainder of the law, amendments, and applications would not be affected. This provision was included because there has been a long history of litigation in this area and is intended to make sure that the entire law would not be overturned if there were a successful legal challenge to one provision.

End Notes

[1] 124 Stat. 1218.

[2] 17 U.S.C. §§ 111, 119, and 122.

[3] 47 U.S.C. §§ 325, 335, 338, 339, 340, and 341.

[4] "Significantly viewed" stations are located outside the local market in which the subscriber is located but have been determined by the Federal Communications Commission to be viewed by a "significant" portion of those households in the local market that do not subscribe to any multichannel video programming distributor (MVPD). The specific threshold viewing level for a significantly viewed station are, for a network affiliate station, a market share of at least 3% of total weekly viewing hours in the market and a net weekly circulation of 25%; for independent stations, 2% of total weekly viewing hours and a net weekly circulation of 5%. The share of viewing hours refers to the total hours that households that do not receive television signals from MVPDs viewed the subject station during the week, expressed as a percentage of the total hours these households viewed all stations during the *week. Net* weekly circulation refers to the number of households that do not receive television signals from MVPDs that viewed the station for 5 minutes or more during the entire week, expressed as a percentage of the total households that do not receive television signals from MVPDs in the survey area. A satellite operator can retransmit the signals of these significantly viewed stations only with the retransmission consent of the station.

[5] The 2004 Satellite Home Viewer Extension and Reauthorization Act allowed satellite operators to retransmit in-state but non-local broadcast television signals to subscribers located in certain counties in Vermont, New Hampshire, Oregon, and Mississippi that are assigned to local markets (as defined by Nielsen Media Research designated market areas) whose local broadcast stations are located in another state. For convenience, these stations are referred to as statutorily exceptional stations.

[6] P.L. 100-667.

[7] P.L. 103-3 69.

[8] P.L. 106-113.

[9] P.L. 108-447, passed as Division J of Title IX of the FY2005 Consolidated Appropriations Act.

[10] For a more detailed discussion of the differences in the rules for cable and satellite providers, see CRS Report R40624, *Reauthorizing the Satellite Home Viewing Provisions in the Communications Act and the Copyright Act: Issues for Congress*, by Charles B. Goldfarb, especially at Table 1, "Current Retransmission and Copyright Rules for Satellite and Cable Operators."

[11] This is formally referred to in the statute as "secondary transmission" of the broadcast signals. The initial transmission of the signals by the broadcast station is the "primary transmission."

[12] *Satellite Home Viewer Extension and Reauthorization Act Section 109 Report*, A Report of the Register of Copyrights, June 2008, at pp. ix-xi and 94-180.

[13] *Satellite Home Viewer Extension and Reauthorization Act Section 109 Report*, A Report of the Register of Copyrights, June 2008, at pp. 167-168.

[14] 47 C.F.R. §§ 76.92, 76.93, 76.106, 76.120, and 76.122. Commercial television station licensees that have contracted with a broadcast network for the exclusive distribution rights to that network's programming within a specified geographic area are entitled to block a local cable system from carrying any programming of a more distant television broadcast station that duplicates that network programming. Commercial broadcast stations may assert these non- duplication rights regardless of whether or not the network programming is actually being retransmitted by the local cable system and regardless of when, or if, the network programming is scheduled to be broadcast. This rule applies to cable systems with more than 1,000 subscribers. Generally, the zone of protection for such programming cannot exceed 35 miles for broadcast stations licensed to a community in the FCC's list of top 100 television markets or 55 miles for broadcast stations licensed to communities in smaller television markets. The non-duplication rule does not apply when the cable system community falls, in whole or in part, within the distant station's Grade B signal contour. In addition, a cable operator does not have to delete the network programming of any station that the FCC has previously recognized as "significantly viewed" in the cable community. With respect to satellite operators, the network non-duplication rule applies only to network signals transmitted by superstations, not to network signals transmitted by other distant network affiliates.

[15] 47 C.F.R. §§ 76.101, 76.103, 76.106, 76. 120, and 76. 123. Cable systems that serve at least 1,000 subscribers may be required, upon proper notification, to provide syndicated protection to broadcasters who have contracted with program suppliers for exclusive exhibition rights to certain programs within specific geographic areas, whether or not the cable system affected is carrying the station requesting this protection. However, no cable system is required to delete a program broadcast by a station that either is significantly viewed in the cable community or places a Grade B or better contour over the community of the cable system. With respect to satellite operators, the syndicated exclusivity rule applies only to syndicated programming transmitted by superstations, not to syndicated programming transmitted by other distant broadcast stations.

[16] 17 U.S.C. §119.

[17] A network is defined as an entity that offers an interconnected program service on a regular basis for 15 or more hours per week to at least 25 affiliated television licensees in 10 or more states. (17 U.S.C. § 1 19(d)(2)(A) and 47 U.S.C. § 339(d)(2)(A)) In addition to the four major television networks—ABC, CBS, Fox, and NBC—that provide national news and entertainment programming aimed at a general audience, there are several networks—Univision, Telefutura, and Telemundo—that offer news and entertainment targeted to ethnic communities, as well as smaller networks that provide entertainment or religious programming to their affiliates. Section 1 19(d)(2)(B) of the Copyright Act defines "network station" to also include noncommercial broadcast stations.

[18] This would include subscribers who are not able to receive network programming because either (1) the satellite operator does not offer the signals of the local broadcast stations and the subscribers are located too far from the transmitter to receive the signals of the local network-affiliated stations over-the-air or (2) there is no network-affiliated station in the local market. The specific household eligibility requirements for receiving distant signals are very complex and include certain grandfathered exceptions, but as a general rule households that can receive the signals of local broadcast television stations either over-the-air or as part of local-into-local satellite service are not eligible to receive distant network signals and would not be affected by the expiration of this provision.

[19] Prior to enactment of STELA, the Copyright Act and the Communications Act both had language referring to "superstations," but that term was defined differently in the two acts, thus creating confusion. The Communications Act identifies a class of "nationally distributed superstations" (47 U.S.C. § 339(d)(2)) that is limited to six stations that were in operation prior to May 1, 1991. These are independent broadcast television stations whose broadcast signals are picked up and redistributed by satellite to local cable television operators and to satellite television operators all across the United States. These nationally distributed superstations in effect function like a cable network rather than a local broadcast television station or a broadcast television network. The nationally distributed superstations are WTBS, Atlanta; WOR and WPIX, New York; WSBK, Boston; WGN, Chicago; KTLA, Los Angeles; and KTVT, Dallas. All of these nationally distributed superstations carry the games of professional sports teams. It has become common in FCC proceedings and discussions to refer to these nationally distributed superstations as simply "superstations." In addition to these independent nationally distributed superstations, there also are many independent television stations that are not nationally distributed superstations. This distinction is important because under section 325(b)(2)(B) of the Communications Act, satellite operators may retransmit the signals of "superstations" without obtaining the consent of the stations if they abide by the FCC's network non-duplication and syndicated exclusivity rules (see footnotes 11 and 12 above), but this exemption from the retransmission consent requirement does apply to the retransmission of the signals of other independent stations. On the other hand, until statutory changes were made in STELA, the Copyright Act had defined "superstation" as "a television station, other than a network station, licensed by the Federal Communications Commission, that is secondarily transmitted by a satellite carrier." (17 U.S.C. § 1 19(d)(9)) Thus, under the Copyright Act pre-STELA, all independent stations were considered superstations and the copyright provisions applied the same way to all independent stations. Language in STELA eliminated the definitional inconsistency between the acts by replacing the word "superstation" with "non-network station" throughout the Copyright Act.

[20] 47 U.S.C. § 325(b)(2)(C).

[21] 47 U.S.C. § 325(b)(3)(C)(ii).

[22] 47 U.S.C. § 325(b)(3)(C)(iii).

[23] Marybeth Peters, Register of Copyrights, written statement before the House Judiciary Committee, hearing on "Copyright Licensing in a Digital Age: Competition, Compensation and the Need to Update the Cable and Satellite TV Licenses," at Appendix 1, February 25, 2009. The proposed modifications to section 111 include revising section 111, and its terms and conditions, to expressly address the retransmission of digital broadcast signals; amending the definition of "local service area of a primary transmitter" to include references to digital station "noise limited service contours" for purposes of defining the local/distant status of noncommercial educational stations (and certain UHF stations) for statutory royalty purposes; amending the statutory definition of "distant signal equivalent" (DSE) to clarify that the royalty payment is for the retransmission of the copyrighted content without regard to the transmission format; amending the definitions of "primary transmission" and "secondary transmission," as well as the "station" definitions in section 111(f) so they comport to the amended definition of DSE; and clarifying that each multicast

stream of a digital television station shall be treated as a separate DSE for section 111 royalty purposes. The proposed modifications to section 119 include replacing the existing Grade B analog standard with the new noise-limited digital signal intensity standard; adopting the Individual Location Longley Rice (ILLR) predictive digital methodology for predicting whether a household can receive an acceptable digital signal from a local digital network station; mandating that the FCC adopt digital signal testing procedures for purposes of determining whether a household is actually unserved by a local digital signal; and deleting various references in section 119 to "analog" unless that reference is to low power television stations that have not yet converted to digital broadcasting.

[24] See, for example, Lauren Lynch Flick and Scott R. Flick, "Congress Passes Satellite Television Extension and Localism Act of 2010," Pillsbury Winthrop Shaw Pittman LLP Client Alert, May 14, 2010, available at *http://www.ilba.org/downloads/~mo~FCC/Congress_Passes _STELA.pdf,* viewed on June 2, 2010. Pillsbury is a law firm with many broadcaster clients.

[25] Warren Communications, *Television & Cable Factbook 2010*, Station Volume 2, "Affiliations by Market for TV Stations, as of October 1, 2009," at pp. C-5 – C-8.

[26] 47 U.S.C. § 339. This provision applies to all network stations, but in practice it primarily involves the retransmission of distant signals into short markets that do not have local broadcast stations affiliated with each of the four major national broadcast networks.

[27] There remains a brief transition period, October 1, 2010, to January 1, 2011, during which if a local broadcaster were to begin multicasting another broadcast network signal, the signal would not be deemed a qualified multicast video and a satellite carrier could import into the local market the signal of a broadcaster affiliated with the same network.

[28] Paul Gallant, an analyst with Stanford Washington Research Group, reportedly stated that mandatory provision of local-into-local service in all markets "would impose significant new costs on Dish Network and DirecTV and generate virtually no new revenue" because the markets in question are so small. See Todd Shields, "DirecTV, Dish May Face Requirement for More Local TV (Update 1)," Bloomberg.com, February 23, 2009, available at http://www.bloomberg.com/apps/news?pid=newsarchive&sid=ayQ_vo3nJImo, viewed on April 27, 2009.

[29] According to the written testimony of Charles W. Ergen, chairman, president, and chief executive officer of DISH Network Corporation, submitted for the hearing on "Reauthorization of the Satellite Home Viewer Extension and Reauthorization Act," before the Subcommittee on Communications, Technology, and the Internet, Committee on Energy and Commerce, U.S. House of Representatives, February 24, 2009, at p. 2, "DISH provides local service in 178 markets today, reaching 97 percent of households nationwide." According to the written testimony of Bob Gabrielli, senior vice president, broadcasting operations and distribution, DIRECTV, Inc., before the House Judiciary Committee, February 25, 2009, at p. 10, "DIRECTV today offers local television stations by satellite in 150 of the 210 local markets in the United States, serving 95 percent of American households. (Along with DISH Network, we offer local service to 98 percent of American households.)"

[30] See John Eggerton, "DISH: Local Into Local Within Two Years—No. 2 DBS Provider Said It Will Deliver Local TV Stations to All 210 DMAs During that Time Frame," *Multichannel News*, October 15, 2009.

[31] *CBS Broad. Inc. v. Echostar Comm. Corp.,* 11th Cir. Docket No. 03-13671 (May 23, 2006).

[32] See, for example, the written testimony of Derek Chang, executive vice president, content strategy and development, DirecTV, Inc., before the House Committee on Energy and Commerce, Subcommittee on Communication Technology, and the Internet, June 16, 2009, at pp. 5-6.

[33] 17 U.S.C. § 119(a)(15)(B).

[34] 17 U.S.C. § 119(a)(15)(D).

[35] 47 U.S.C. § 335(b)(1).

[36] Written Testimony of Bill Acker, Director of Broadcasting and Technology, West Virginia Public Broadcasting, before the Senate Committee on Commerce, Science, and Transportation, Subcommittee on Communications, Technology and the Internet, October 7, 2009, at p. 3.

[37] Ibid. at pp. 8-10.

[38] The statutory provisions for satellite explicitly require the use of Nielsen's DMAs. (17 U.S.C. § 122(j)(2)(A) and (C).) The statutory provisions for cable instructed the FCC to make market determinations "using, where available, commercial publications which delineate television markets based on viewing patterns." (47 U.S.C. § 534(h)(1)(C).) Nielsen had already delineated such television markets, assigning geographic areas to markets based on predominant viewing patterns in order to construct ratings data for advertisers, and the FCC therefore adopted Nielsen's market delineations.

[39] For a complete state-by-state list of these counties, their populations, and the full power television stations located in the counties, see the Appendix to CRS Report R40624, *Reauthorizing the Satellite Home Viewing Provisions in the Communications Act and the Copyright Act: Issues for Congress*, by Charles B. Goldfarb.

[40] 17 U.S.C. §§ 119(a)(2)(c)(i)-(iv) and 47 U.S.C. § 341.

[41] See, for example, John Eggerton, "Affiliate Associations Warn Legislators Against Allowing Imported Signals from In-State, Distant Markets," *Broadcasting & Cable,* March 30, 2009.

[42] Representative Ross has introduced H.R. 3216, the Local Television Freedom Act of 2009, which would allow multichannel video programming distributors (MVPDs)—satellite operators and cable operators (including telephone companies)—serving an orphan county to retransmit to their subscribers in that county the signals of television broadcast stations located in an adjacent in-state market. In addition, the Four Corners Television Access Act of 2009 has been introduced in both the House (H.R. 1860, by Representatives Salazar and Coffman) and the Senate (S. 771, by Senators Bennet and Udall) to allow satellite operators to retransmit the signals of certain in-state broadcast stations to subscribers located in two Colorado counties that are assigned to the Albuquerque, NM, local market and to allow cable operators located in those counties to retransmit the signals of certain in-state stations without having to obtain retransmission consent from the stations. Also, Representative Boren has introduced H.R. 505, which would allow satellite operators to retransmit to any subscriber in the state of Oklahoma—not just those in orphan counties—the signals of any broadcast station located in that state.

[43] See Anandashankar Mazundar, "Senate Judiciary Committee Votes Out Satellite Television Reauthorization Bill," *BNA Daily Report for Executives*, September 25, 2009.

[44] *Satellite Home Viewer Extension and Reauthorization Act Section 109 Report*, A Report of the Register of Copyrights, June 2008, at p. xiv.

[45] See, for example, the Written Testimony of Robert Gabrielli, senior vice president for program operations, DirecTV, Inc., before the Senate Committee on Commerce, Science, and Transportation, October 7, 2009, at p. 8.

[46] See, for example, the website of Public Knowledge at *http://www.publicknowledge.org/issues/copyright*

[47] Prior to this clarification, there have been situations in which a cable operator has been required to make a copyright payment as if it were retransmitting a distant signal to all the communities in a service area, but in fact was not allowed to retransmit the signal to certain communities in the service area. Cable operators have referred to the signals that they were not allowed to retransmit, but for which they had to make copyright payments, as "phantom signals."

In: Media Industry Programming, Competition...   ISBN: 978-1-61122-078-0
Editors: Ryan E. Moore   ©2011 Nova Science Publishers, Inc.

Chapter 3

LEGAL CHALLENGE TO THE FCC'S MEDIA OWNERSHIP RULES: AN OVERVIEW OF *PROMETHEUS RADIO V. FCC* AND RECENT REGULATORY DEVELOPMENTS

Kathleen Ann Ruane

SUMMARY

In December 2007, the Federal Communications Commission relaxed its newspaper/broadcast ownership ban (order released February 2008). The decision raised concerns in Congress about increasing media consolidation that have long been at the forefront of the debate over ownership restrictions. The Commission's order served to rekindle the discussion of media consolidation and the perceived need to take action to preserve a diversity of voices in the marketplace of ideas. The FCC rule, as this report illustrates, has a history dating back to a previous failed attempt to relax a greater number of broadcast cross-ownership restrictions, and it is worthwhile to examine this previous proceeding in order to understand the current status of the rules.

On June 2, 2003, the FCC adopted a set of comprehensive rules addressing six different aspects of media ownership, including cross-ownership of broadcast and print media, local television and radio ownership,

and national television ownership. On June 24, 2004, the United States Court
of Appeals for the Third Circuit, in *Prometheus Radio v. FCC*, remanded
several of these rules to the Commission for further consideration finding that
the Commission failed to adequately justify the numerical limitations used in
the rules. This report provides an overview of the Commission's 2002 Biennial
Review from which the 2003 rules originated and the *Prometheus* case.

The report also addresses current issues facing the actions taken by the
FCC in response to the Third Circuit Court of Appeals' decision in
Prometheus. On December 18, 2007, the FCC concluded its review of
broadcast ownership rules by relaxing the newspaper/broadcast station cross-
ownership restrictions in certain markets. All other broadcast ownership rules,
however, remain unchanged.

The relaxation of the newspaper/broadcast cross-ownership rule as well as
the other ownership rules passed by the FCC in December 2007 have yet to go
into effect. Pursuant to the Third Circuit's final order in the *Prometheus* case,
the FCC's newest rules may not take effect until the Third Circuit lifts its stay.
On June 12, 2009, the Third Circuit decided to keep the stay in place until
further order of the court. On October 1, 2009, the FCC filed a status report
with the Third Circuit. The FCC argued that the stay should remain in place,
because the 2008 order no longer incorporates the views of a majority of the
Commissioners and the agency is set to begin a new review of the media
ownership rules that should be completed in 2010.

INTRODUCTION

On December 18, 2007, the Federal Communications Commission
("FCC" or "Commission") concluded a review of its broadcast ownership
rules by relaxing the ban on cross-ownership of a newspaper and a broadcast
station in certain markets.[1] The order adopted that day ended agency
proceedings that had been ongoing for five years.[2] In 2003, the FCC had
adopted a comprehensive order (in its 2002 Biennial Review) revising many of
its cross-ownership rules but, as will be discussed below, the United States
Court of Appeals for the Third Circuit found insufficient basis for many of the
proposed changes in that order and remanded it to the FCC for
reconsideration. This report discusses the 2002 Biennial Review, the decision
by the Third Circuit that struck many of those rules down, and the FCC's
actions upon remand. The report also addresses the current status of the rules.

The Telecommunications Act of 1996 sought to create a "pro-competitive, deregulatory national policy framework designed to accelerate rapidly private sector development of advanced telecommunications and information technologies and services to all Americans by opening all telecommunications markets to competition."[3] Among other things, the act eliminated limits on national radio ownership, raised the cap on the percentage of the national audience that a single station owner may reach, set new limits for local radio ownership, and directed the Commission to conduct a rulemaking proceeding to determine whether to retain, modify, or eliminate the local television ownership limitations.[4] The act also directed the Commission to review its broadcast ownership rules every two years to "determine whether any of such rules are necessary in the public interest as the result of competition."[5]

The Commission initiated its 2002 Biennial Review in September of 2002 with a Notice of Proposed Rulemaking announcing that it would review four of its broadcast ownership rules: the national audience reach limit; the local television rule; the radio/television cross-ownership ("one-to-a-market") rule; and the dual network ownership rule.[6] The Commission had previously initiated proceedings regarding the local radio ownership rule and the newspaper/broadcast cross- ownership rule.[7] Those proceedings were incorporated into the Biennial Review.

On June 2, 2003, the Commission adopted a Report and Order modifying its ownership rules.[8] In the Order, the Commission concluded that "neither an absolute prohibition on common ownership of daily newspapers and broadcast outlets in the same market (the 'newspaper/broadcast cross- ownership rule') nor a cross-service restriction on common ownership of radio and television outlets in the same market (the 'radio-television cross-ownership rule') [remained] necessary in the public interest."[9] The Commission found that "the ends sought can be achieved with more precision and with greater deference to First Amendment interests through [its] modified Cross Media Limits ('CML')."[10] The Commission also revised the market definition and the way it counted stations for purposes of the local radio rule, revised the local television multiple ownership rule to permit the common ownership of up to three stations in large markets, modified the national television ownership cap to raise the national audience reach limit to 45%, and retained the dual network rule.

Following the publication of the Commission's Order, several organizations filed petitions for review of the new rules. The petitions for review were consolidated and heard by the United States Court of Appeals for the Third Circuit. After an initial hearing on September 3, 2003, the court entered a stay for the effective date of the proposed rules, preventing their enforcement, and ordered that the prior ownership rules remain in effect pending resolution of the proceedings.[11] On February 14, 2004, the court heard oral arguments and issued its opinion on June 24, 2004.[12]

2002 BIENNIAL REVIEW

As noted above, on June 2, 2003, the Commission approved a Report and Order modifying its media ownership rules to provide a "new, comprehensive framework for broadcast ownership regulation."[13] The Commission determined that new technologies necessitated new rules and that the prior rules "inadequately [accounted] for the competitive presence of cable, [ignored] the diversity-enhancing value of the Internet, and [lacked] any sound basis for a national audience reach cap."[14] According to the Commission, the newly adopted rules were "not blind to the world around them, but reflective of it," and "necessary in the public interest."[15]

National Ownership Rules

With respect to the ownership of broadcast stations on a nationwide level, the Commission determined that while "a national TV ownership limit is necessary to promote localism by preserving the bargaining power of affiliates and ensuring their ability to select programming responsive to tastes and needs of their local communities," the evidence demonstrated that a 35% cap was not necessary to "preserve that balance" and raised the limit to 45%.[16] Under the new rule, a single entity was prohibited from owning stations that would allow it to reach more than 45% of the national audience. The Commission also elected to retain the "UHF discount," which attributes UHF stations with only 50% of the households in their DMA, despite many cable operators now carrying UHF stations.

While it modified the national television ownership cap, the Commission determined that its dual network rule, which prohibits common ownership of the top four television networks, remained necessary in the public interest and did not attempt to repeal or modify it.[17]

Local Ownership Rules

In the 2002 Biennial Review, the Commission either modified or repealed its local ownership rules. The cross-ownership rules prohibiting the common ownership of a full-service broadcast television station and a daily newspaper in the same community and limiting the ownership of television and radio combinations by a single entity in a given market were both repealed.[18] The Commission determined that neither rule remained necessary in the public interest and replaced both rules with a single set of cross-media limits based on market size. In large markets, defined as those with more than eight television stations, cross-ownership was unrestricted.

The Commission combined an earlier remand from the D.C. Circuit Court of Appeals[19] of its modified "duopoly rule" with the 2002 Biennial Review and adopted a new rule that would permit common ownership of two commercial television stations in markets that have seventeen or fewer full-power commercial and noncommercial stations, and common ownership of three commercial stations in markets that have eighteen or more stations.[20] These limitations are subject to a further restriction on the common ownership of stations that are ranked among the market's largest four stations based on audience share. The Commission also elected to repeal the "Failed Station Solicitation Rule" related to the sale of failed, failing, or unbuilt stations, which required notice of the sale to be provided to out-of-market buyers.

With respect to local radio ownership, the FCC modified its prior rule by adopting a new method for determining the size of a local market, but retaining the rule's prior numerical limits on station ownership.[21] The Commission's prior regulations defined the local market by using the "contour-overlap methodology,"[22] which the Commission abandoned in favor of the "geography-based market definition used by Arbitron, a private entity that measures local radio audiences for its customer stations."[23] The Arbitron markets include both commercial and noncommercial stations. While it changed the definition of local market, the Commission retained its numerical limits, which allow a single entity to own as many as eight radio stations in markets of 45 or more commercial stations.[24]

An additional modification to the local radio ownership rule created a new system for the attribution of joint sales agreements (JSAs).[25] Generally, a JSA authorizes a broker to sell advertising time for the brokered station in return for a fee paid to the licensee. The Commission noted that because the broker station normally assumes much of the market risk with respect to the station it brokers, it typically has the authority to make decisions with respect to the sale of advertising time on the station. Under the prior rules, JSAs were not attributable to the brokering entity and were not counted toward the number of stations the brokering licensee may own in a local market. The new rules made the JSAs attributable to the brokering entity for the purpose of determining the brokering entity's compliance with the local ownership limits if the brokering entity owns or has an attributable interest in one or more stations in the local market, and the joint advertising sales amount to more than 15% of the brokered station's advertising time per week.

THE COURT'S DECISION

Several organizations filed petitions for review of the new rules upon their publication. The numerous petitions for review were consolidated and the case was heard by the United States Court of Appeals for the Third Circuit in Philadelphia. As noted above, after an initial hearing, the court entered a stay for the effective date of the proposed rules.[26] On February 14, 2004, the court heard oral arguments and issued its opinion on June 24, 2004.[27]

With respect to the national ownership rules, the court did not address the Commission's decision to raise the national audience reach cap from 35% to 45% citing Congress's modification of the rule in the 2004 Consolidated Appropriations Act.[28] Section 629 of the act directed the Commission to modify the rule by setting a 39% cap on national audience reach.[29] The court determined that because the Commission was under "a statutory directive to modify the national television ownership cap to 39%, challenges to the Commission's decision to raise the cap to 45 were moot."[30]

Additional challenges to the UHF discount provisions in the rule were also deemed moot even though the UHF discount rules were not mentioned in the 2004 Consolidated Appropriations Act. The court determined that the UHF discount was intrinsically linked to the 39% national audience reach cap because "reducing or eliminating the discount for UHF stations audiences would effectively raise the audience reach limit."[31] The court also noted with respect to the UHF discount that the 2004 Consolidated Appropriations Act specifically provided that the periodic review provisions set forth in the amendment did not apply to "any rules relating to the 39% national audience limitation," and as a rule "relating to" the national audience limitation, Congress intended to insulate the UHF discount from review.

None of the parties bringing the *Prometheus* case challenged the retention of the dual network rule, so this was not addressed by the court.

With respect to the Commission's local ownership rules, the court agreed with the Commission's decision to modify these rules in many respects. However, the court found fault with the numerical limits set by the FCC in each of the local ownership rules. The court stated that "[t]he Commission's derivation of new Cross-Media Limits, and its modification of the numerical limits on both television and radio station ownership in local markets, all have the same essential flaw: an unjustified assumption that media outlets of the same type make an equal contribution to diversity and competition in local markets."[32]

The court determined that the Commission's decision to repeal the ban on broadcast/newspaper cross-ownership was justified and supported by evidence in the record and found that the Commission's decision to retain some limits on common ownership was constitutional and not in violation of the Communications Act.[33] However, the court found that the FCC failed to provide reasoned analysis to support the specific limits that it chose with respect to the new "cross-media" rules, stating that the limits "employ several irrational assumptions and inconsistencies."[34] The court rejected the Commission's use of a "diversity index,"[35] because of what the court saw as the fallacies upon which it was based and because the Commission failed to provide adequate notice of the new methodology in the rulemaking proceedings leading up to the 2002 Order.[36] The court remanded the cross-media limits and advised the Commission to make any "new metric for measuring diversity and competition in a market ... subject to public notice and comment before it is incorporated into a final rule."[37]

The court in *Prometheus* upheld the restriction on common ownership of the market's top four broadcast television stations, but remanded the numerical limits "for the Commission to harmonize certain inconsistencies and better support its assumptions and rationale."[38] In making its decision, the court found that the Commission had presented evidence in the record to adequately support the "top-four restriction,"[39] while failing to justify the market share assumptions used as the basis for the numerical limits. The court stated that "[n]o evidence supports the Commission's equal market share assumption, and no reasonable explanation underlies its decision to disregard actual market share."[40] The court also remanded the Commission's repeal of the Failed Station Solicitation Rule, finding that the Commission failed to consider "the effect of its decision on minority television station ownership," and thus failed "'to consider an important aspect of the problem' [amounting] to arbitrary and capricious rulemaking."[41]

In addition to upholding the Commission's restriction on common ownership of a market's top four broadcast television stations, the court upheld the Commission's new definition of local markets with respect to radio finding that the Commission's decision was "in the public interest" and that it was a "rational exercise of rulemaking authority."[42] The court also found that the Commission justified the inclusion of noncommercial stations in the new definition. However, with respect to the numerical limits retained by the Commission, the court concluded that while the numerical limits approach was rational and in the public interest, the Commission failed to support its decision to retain these particular limits with "reasoned analysis."[43] The court rejected the Commission's contention that five equal-sized competitors would ensure that local markets are competitive, and found that even if it were to justify the "five equal-sized competitors" benchmark, that it failed to sufficiently demonstrate that under the existing numerical limits five equal-sized competitors would actually emerge.[44] The court remanded the numerical limits to the Commission "to develop numerical limits that are supported by a rational analysis."[45]

With respect to the new rules providing for the attribution of joint sales agreements, the court affirmed the Commission's decision, finding that the Commission changed its rules as the result of "reasoned decisionmaking," and that such a change was "necessary in the public interest" due to "the potential for brokering entities to influence the brokered stations."[46]

POST-*PROMETHEUS*

On September 3, 2004, the Third Circuit granted the Commission's motion requesting a partial lifting of the stay to allow those parts of the rules approved by the court in its June 24 decision to go into effect. Specifically, the stay was lifted with respect to the use of Arbitron metro markets to define local markets, the inclusion of noncommercial stations in determining the size of a market, the attribution of stations whose advertising is brokered under a Joint Sales Agreement to a brokering station's permissible ownership totals, and the imposition of a transfer restriction. The stay remained in place pending FCC action on remand with respect to all other aspects of the Biennial Review Order.[47]

On January 27, 2005, the United States Solicitor General and the FCC decided not to appeal the Third Circuit's decision.[48] However, several media companies filed a formal appeal with the Supreme Court asking for a review of the Third Circuit's decision.[49] On June 13, 2005, the Supreme Court denied certiorari in all relevant appeals.[50]

2007 BROADCAST OWNERSHIP RULES

On July 24, 2006, the FCC issued a Further Notice of Proposed Rulemaking (FNPR) in the Broadcast Media Ownership proceedings that had been remanded to the Commission in 2003.[51] The FNPR sought comment for new ownership rules that would comport with the Third Circuit's decision in *Prometheus*.[52] Specifically, the FCC sought comment suggesting new rules that would foster "localism;" increase opportunities for ownership among minorities and women; revise the numerical limits placed on cross ownership of local television stations and local radio stations; revise the Diversity Index used to calculate the availability of outlets that contribute to diversity of viewpoints in local media markets; and other suggestions for improvement of existing and proposed rules.[53] The FCC also commissioned multiple studies on media ownership and sought comment on these studies to determine whether and to what extent to take the studies into account in the final ownership rules.[54] The reply comment period on the ownership studies closed November 1, 2007.[55]

On August 1, 2007, the FCC issued a Second Further Notice of Proposed Rulemaking (SFNPR) in its ongoing review of the broadcast ownership rules.[56] The SFNPR sought comments on new initiatives specifically related to encouraging minority and female ownership of broadcast stations proposed by the Minority Media and Telecommunications Council (MMTC), as well as potential constitutional issues related to race specific classifications.[57] Reply comments were due for the SFNPR on October 16, 2007.[58]

On November 13, 2007, following the close of all comment and reply comment periods, FCC Chairman Martin proposed that the review of broadcast ownership rules should conclude by adopting a relaxation of the ban on newspaper and broadcast cross-ownership.[59] The proposal also indicated that no changes would be made in the local television "duopoly" rule, the local radio ownership rule, or the local radio-television cross-ownership rule already in force.

The FCC adopted a revised version of Chairman Martin's proposal to ease the ban on newspaper/broadcast cross-ownership on December 18, 2007.[60] The Report and Order in the proceeding was released on February 4, 2008.[61] The new rule establishes the presumption that newspaper/radio broadcast station cross-ownership in the top 20 largest DMAs is in the public interest, and that newspaper/television broadcast station cross-ownership in the top 20 largest DMAs is in the public interest when the television station is not among the top four ranked stations in the market and at least eight "major media voices" would remain in the DMA post- merger. [62] For all other DMAs, the new rule establishes the presumption that newspaper/broadcast station cross-ownership is not in the public interest, except in two circumstances (discussed below).[63] Applicants attempting to overcome a presumption that the proposed combination is not in the public interest will have to demonstrate, through clear and convincing evidence, that the merged entity will increase the diversity of independent news outlets and increase competition among independent news sources in the relevant market.[64] The FCC also has laid out four factors to help inform its evaluation of these proposed combinations.[65]

The new rules identify two circumstances in which the presumption that cross-ownership is not in the public interest will be reversed.[66] The first circumstance adapts the FCC's failed or failing station waivers to newspaper/broadcast combinations.[67] Therefore, when either the broadcast station or the newspaper involved in a proposed combination is "failed" or "failing," the FCC will presume that the proposed combination is in the public interest.[68] The presumption that a combination is not in the public interest also will be reversed when the proposed combination will result in a new source of local news in a market, specifically defined as a combination that would initiate at least seven hours of new local news programming per week on a broadcast station that previously has not aired local news.[69] All other cross-ownership rules and restrictions will remain unchanged. [70]

The FCC also adopted rules in December 2007 to promote diversification of broadcast ownership in a separate order from the newspaper/broadcast station cross-ownership rule. The new rules are intended to allow "eligible entities" to more easily access financing and spectrum by, for example, modifying the distress sale policy to allow a licensee whose licenses were designated for a revocation hearing to sell its station to an eligible entity prior to the commencement of the hearing, revising the FCC's equity/debt plus attribution standard to facilitate investment in eligible entities, and giving priority to any entity financing an eligible entity in certain duopoly situations.[71] "Eligible entities" are defined as "entities that would qualify as a small business consistent with Small Business Administration standards, based on revenue."[72] The FCC is seeking further comment on whether it can expand the definition of "eligible entity" to include other business.[73]

RECENT COURT PROCEEDINGS

The relaxation of the newspaper/broadcast cross-ownership rule as well as the other ownership rules promulgated by the FCC in December 2007 have yet to go into effect. Pursuant to the Third Circuit's final order in the *Prometheus* case, the FCC's newest rules may not go into effect until the Third Circuit lifts its stay.[74] On June 12, 2009, the Third Circuit decided to keep the stay in place until further order of the court and ordered the parties to file status reports regarding whether the stay should remain in place later in the year.[75]

On October 1, 2009, the FCC filed its status report with the Third Circuit. The FCC argued that the stay should remain in place, because the 2008 order no longer incorporates the views of a majority of the Commissioners and the agency is set to begin a new review of the media ownership rules that should be completed in 2010.[76]

On December 18, 2009, the Third Circuit ordered the FCC to show cause as to why the court's stay on the newspaper/broadcast cross-ownership rule changes should not be lifted.[77] The FCC filed its brief on the issue with the court on January 7.[78] The parties await the court's decision.

ACKNOWLEDGMENTS

This report was originally written by Angie A. Welborn, Legislative Attorney.

End Notes

[1] *In the Matter of 2006 Quadrennial Regulatory Review—Review of the Commission's Broadcast Ownership Rules and Other Rules Adopted Pursuant to Section 202 of the Telecommunications Act of 1996; 2002 Biennial Regulatory Review—Review of the Commission's Broadcast Ownership Rules Pursuant to Section 202 of the Telecommunications Act of 1996; Cross-Ownership of Broadcast Stations and Newspapers; Rules and Policies Concerning Multiple Ownership of Radio Broadcast Stations in Local Markets; Definition of Radio Markets; Ways to Further Section 257 Mandate to Build on*

Earlier Studies; Public Interest Obligations of TV Broadcast Licensees, MB Docket No. 06-121, MB Docket No. 02-227, MM Docket No. 0 1-235, MM Docket No. 01-317, MM Docket No. 00-244, MB Docket No. 04-228, MM Docket No. 99-3 60 (Released February 4, 2008), 2008 FCC LEXIS 1083.

[2] *Id.* The FCC consolidated the proceeding remanded by the Third Circuit in the *Prometheus* case with its quadrennial review of its broadcast ownership rules and other broadcast ownership proceedings.

[3] S.Rept. 104-230, pp. 1-2 (1996).

[4] Telecommunications Act of 1996, P.L. 104-104 (1996).

[5] P.L. 104-104, Sec. 202(h).

[6] *In the Matter of 2002 Biennial Regulatory Review—Review of the Commission's Broadcast Ownership Rules and Other Rules Adopted Pursuant to Section 202 of the Telecommunications Act of 1996*, 17 FCC Rcd 18503 (2002).

[7] *See* 16 FCC Rcd 19861 (2001) and 16 FCC Rcd 17283 (2001).

[8] *In the Matter of 2002 Biennial Regulatory Review*, Report and Order and Notice of Proposed Rulemaking, 18 FCC Rcd 13620 (2003). Hereinafter, cited as Report and Order. For more information on the Commission's media ownership rules, see CRS Report RL34416, *The FCC's Broadcast Media Ownership Rules*, by Charles B. Goldfarb.

[9] *Id* at ¶ 2.

[10] *Id.*

[11] *Prometheus Radio Project v. FCC*, 2003 U.S. App. LEXIS 18390 (3rd Cir. 2003).

[12] *Prometheus Radio Project v. FCC*, 373 F.3d 372 (3rd Cir. 2004).

[13] Report and Order, ¶ 3.

[14] *Id.* at ¶ 4.

[15] *Id.* at ¶ 8.

[16] *Id.* at ¶ 507.

[17] *Id.* at ¶ 592.

[18] *Id.* at ¶ 327.

[19] *Sinclair Broadcast Group v. FCC*, 284 F.3d 148 (D.C. Cir. 2002). The Telecommunications Act of 1996 directed the FCC to determine whether to "retain, modify, or eliminate its limitations on the number of television stations that a person or entity may own, operate, or control, or have a congnizable interest in, within the same television market." P.L. 104-104, Sec. 202(c)(2). In response to this directive, the Commission modified its rules in 2000 to allow an entity to own two television stations in a DMA (designated market area), provided that (1) the Grade B field strength contours of the stations do not overlap; and (2) at least one of the stations is not ranked among the top four highest-ranked stations in the DMA, and at least eight "independent voices" would remain in the DMA after the proposed combination. The United States Court of Appeals for the D.C. Circuit reviewed this rule, and remanded it to the Commission to justify its definition of "voices," which included only broadcast television stations and not other types of non-broadcast media. The Commission consolidated the Sinclair remand with its 2002 Biennial Review leading to this challenge.

[20] Report and Order, ¶ 186.

[21] *Id.* at ¶ 235 et seq. For more information, CRS Report RL3 1925, *FCC Media Ownership Rules: Current Status and Issues for Congress*, by Charles B. Goldfarb.

[22] For a description of the "contour-overlap methodology," see *supra note* 6 at Appendix F.

[23] The Telecommunications Act of 1996 did not define local markets using any particular methodology.

[24] The Telecommunications Act of 1996 established the current numerical limits. Under the '96 Act, in a radio market with 45 or more commercial radio stations, a party may own, operate, or control up to 8 commercial radio stations, not more than 5 of which are in the same service (AM or FM); in a radio market with between 30 and 44 (inclusive) commercial radio stations, a party may own, operate, or control up to 7 commercial radio stations, not more than 4 of which are in the same service (AM or FM); in a radio market with between

15 and 29 (inclusive) commercial radio stations, a party may own, operate, or control up to 6 commercial radio stations, not more than 4 of which are in the same service (AM or FM); and in a radio market with 14 or fewer commercial radio stations, a party may own, operate, or control up to 5 commercial radio stations, not more than 3 of which are in the same service (AM or FM), except that a party may not own, operate, or control more than 50 percent of the stations in such market. P.L. 104-104, Sec. 202(b).

[25] Report and Order, ¶ 317.
[26] *Prometheus Radio Project v. FCC,* 2003 U.S. App. LEXIS 18390 (3rd Cir. 2003).
[27] *Prometheus Radio Project v. FCC,* 373 F.3d 372 (3rd Cir. 2004).
[28] P.L. 108-199, Sec. 629.
[29] Section 629 also amended section 202(h) of the Telecommunications Act of 1996 to change the review period from a biennial review to a quadrennial review, and it exempted the 39% cap on national audience reach from review.
[30] *Prometheus,* 373 F.3d at 396.
[31] *Id.*
[32] *Id.* at 435.
[33] *Id.* at 397 - 401.
[34] *Id.* at 402.
[35] The Commission's diversity index was not based on the actual market shares of companies, but rather on the assumption that each television station in a market provides the same diversity impact regardless of the actual size of its audience, and the same for each newspaper, each radio station, etc. The court rejected the contention that each outlet provides the same diversity impact, saying that "[a] diversity index that requires us to accept that a community college television station makes a greater contribution to viewpoint diversity than a conglomerate that includes the third-largest newspaper in America also requires us to abandon both logic and reality." *Prometheus,* 373 F.3d at 408.
[36] *Id.* at 411-413.
[37] *Id.* at 412.
[38] *Id.* at 412.
[39] *Id.* at 418.
[40] *Id.* at 420.
[41] *Id.* at 421.
[42] *Id.* at 425.
[43] *Id.* at 426.
[44] *Id.* at 432-433.
[45] *Id.* at 434.
[46] *Id.* at 429-430.
[47] *Prometheus Radio v. FCC,* 03-3388 (3rd Cir., September 3, 2004).
[48] *Feds Leave Broadcasters Alone in FCC Media Ownership Appeal,* Communications Daily, January 28, 2005.
[49] *Media Group Asks Supreme Court to Hear Ownership Case,* Communications Daily, January 31, 2005.
[50] *Media Gen., Inc. v. FCC,* 2005 U.S. LEXIS 4807 (June 13, 2005).
[51] *In the Matter of 2006 Quadrennial Regulatory Review - Review of the Commission's Broadcast Ownership Rules and Other Rules Adopted Pursuant to Section 202 of the Telecommunications Act of 1996,* Further Notice of Proposed Rulemaking, 21 FCC Rcd 8834 (July 24, 2006).
[52] *Id.*
[53] *Id.* For a thorough discussion of the rules proposed in 2002 and the current state of the FCC's media ownership rules, see CRS Report RL34416, *The FCC's Broadcast Media Ownership Rules,* by Charles B. Goldfarb.

[54] FCC Seeks Comment on Research Studies on Media Ownership, Public Notice, MB Docket No. 06-121 (July 31, 2007), *available at http://fjallfoss.fcc.gov/edocs_public/attachmatch/ DA-07-3470A1.pdf.*

[55] Media Bureau Extends Filing Deadline for Comments on Media Ownership Studies, Public Notice, MB Docket No. 06-121 (September 28, 2007), *available at http://fjallfoss.fcc.gov/ edocs_public/attachmatch/DA-07-4097A1.pdf.*

[56] *In the Matter of 2006 Quadrennial Regulatory Review - Review of the Commission's Broadcast Ownership Rules and Other Rules Adopted Pursuant to Section 202 of the Telecommunications Act of 1996*, Second Further Notice of Proposed Rule Making, 2007 FCC LEXIS 5775 (August 1, 2007).

[57] *Id.*

[58] *In the Matter of 2006 Quadrennial Regulatory Review - Review of the Commission's Broadcast Ownership Rules and Other Rules Adopted Pursuant to Section 202 of the Telecommunications Act of 1996*, Second Further Notice of Proposed Rule Making, 2007 FCC LEXIS 5775 (August 1, 2007).

[59] Press Release, Federal Communications Commission, Chairman Kevin J. Martin Proposes Revision to Newspaper/Broadcast Cross-Ownership Rule (November 13, 2007), available at http://hraunfoss.fcc.gov/edocs_public/ attachmatch/DOC-278 1 13A1 .pdf.

[60] Press Release, Federal Communications Commission, FCC Adopts Revision to Newspaper/Broadcast Cross- Ownership Rule (December 18, 2007), available at http://hraunfoss.fcc.gov/edocs_public/attachmatch/DOC278932A1 .pdf.

[61] *In the Matter of 2006 Quadrennial Regulatory Review—Review of the Commission's Broadcast Ownership Rules and Other Rules Adopted Pursuant to Section 202 of the Telecommunications Act of 1996; 2002 Biennial Regulatory Review—Review of the Commission's Broadcast Ownership Rules Pursuant to Section 202 of the Telecommunications Act of 1996; Cross-Ownership of Broadcast Stations and Newspapers; Rules and Policies Concerning Multiple Ownership of Radio Broadcast Stations in Local Markets; Definition of Radio Markets; Ways to Further Section 257 Mandate to Build on Earlier Studies; Public Interest Obligations of TV Broadcast Licensees*, MB Docket No. 06-121, MB Docket No. 02-227, MM Docket No. 0 1-235, MM Docket No. 01-317, MM Docket No. 00-244, MB Docket No. 04-228, MM Docket No. 99-360 (Released February 4, 2008), 23 FCC Rcd 2010.

[62] *Id.* at ¶¶ 20, 53-62.

[63] *Id.* at ¶¶ 20, 63-75.

[64] *Id.* at ¶ 68.

[65] *Id.*

[66] *Id.* at ¶ 65.

[67] *Id.* at ¶¶ 65-66.

[68] *Id.*

[69] *Id.* at ¶ 67.

[70] *Id.* at ¶ 1.

[71] *In the Matter of Promoting Diversification of Ownership in Broadcasting Services, 2006 Quadrennial Regulatory Review—Review of the Commission's Broadcast Ownership Rules and Other Rules Adopted Pursuant to Section 202 of the Telecommunications Act of 1996, 2002 Regulatory Review—Review of the Commission's Broadcast Ownership Rules and Other Rules Adopted Pursuant to Section 202 of the Telecommunications Act of 1996, Cross-Ownership of Broadcast Stations and Newspapers, Rules and Policies Concerning Multiple Ownership of Radio Broadcast Stations in Local Markets, Definition of Radio Markets, Ways to Further Section 257 Mandate to Build on Earlier Studies*, MB Docket No. 07-294, MB Docket No. 06-121, MB Docket No. 02-277, MM docket No. 01-235, MM Docket No. 01- 317, MM Docket No. 00-244, MB Docket No. 04-228 adopted December 18, 2007, released March 5, 2008.

[72] *Id.*

[73] *Id.*

[74] *Prometheus*, 373 F.3d at 435 ("The stay currently in effect will continue pending our review of the Commission's action on remand, over which this panel retains jurisdiction.").

[75] Order Continuing Stay, *Prometheus Radio Project v. FCC*, Nos. 08-9078 et al. (June 12, 2009). *See* also John Eggerton, *Court Won't Lift Stay on Newspaper/Broadcast Cross-Ownership Rule Change*, BROADCASTING & CABLE, June 12, 2009, available at http://www.broadcastingcable.com/article/279319-Court_Won_t_Lift_Stay_On_Newspaper_Broadcast_Crossownership_Rule_Change.php.

[76] Status Report of the Federal Communications Commission, *Prometheus Radio Project v. FCC*, No. 08-3078 et al. (October 1, 2009) available at *http://hraunfoss.fcc.gov/edocs_public/attachmatch/DOC*

[77] John Eggerton, Third Circuit to Lift Stay on FCC's Cross-Ownership Rule Revision, *Multichannel News*, December 18, 2009, http://www.multichannel.com/article/441161-Third_Circuit_to_Lift_Stay_On_FCC_s_Cross_Ownership_Rule_Revision.php.

[78] John Eggerton, FCC Defends Cross-Ownership Stay, *Multichannel News*, January 7, 2010, http://www.multichannel.com/article/443122-FCC_Defends_Crossownership_Stay.php?rssid=20062.

In: Media Industry Programming, Competition...   ISBN: 978-1-61122-078-0
Editors: Ryan E. Moore   ©2011 Nova Science Publishers, Inc.

Chapter 4

JOINT WRITTEN TESTIMONY OF BRIAN L. ROBERTS, CHAIRMAN AND CHIEF EXECUTIVE OFFICER, COMCAST CORPORATION AND JEFF ZUCKER, PRESIDENT AND CHIEF EXECUTIVE OFFICER, NBC UNIVERSAL, BEFORE THE COMMITTEE TO THE JUDICIARY, HEARING ON "COMPETITION IN THE MEDIA AND ENTERTAINMENT DISTRIBUTION MARKET"

Brian L. Roberts

Mr. Chairman, and Members of the Committee, we are pleased to appear before you today to discuss Comcast Corporation's ("Comcast") planned joint venture with General Electric Company ("GE"), under which Comcast will acquire a majority interest in and management of NBC Universal ("NBCU"). As you know, the proposed transaction will combine in a new joint venture the broadcast, cable programming, movie studio, theme park, and online content businesses of NBCU with the cable programming and certain online content businesses of Comcast. This content-focused joint venture will retain the NBCU name.

The new NBCU will benefit consumers and will encourage much-needed investment and innovation in the important media sector.

How will it benefit consumers?

First, the new venture will lead to increased investment in NBCU by putting these important content assets under the control of a company that is focused exclusively on the communication and entertainment industry. This will foster enhanced investment in both content development and delivery, enabling the new NBCU to become a more competitive and innovative player in the turbulent and ever-changing media world. Investment and innovation will also preserve and create sustainable media and technology jobs in the U.S.

Second, the transaction will promote the innovation, content, and delivery that consumers want and demand. The parties have made significant commitments in the areas of local news and information programming, enhanced programming for diverse audiences, and more quality educational and other content for children and families.

And finally, Comcast's commitment to preserve NBCU's journalistic independence and to sustain and invest in the NBC broadcast network will promote the quality news, sports, and diverse programming that have made this network great over the last 50 years. We discuss these specific and verifiable public interest commitments later in this testimony, and a summary is attached.

The new NBCU will advance key policy goals of Congress: diversity, localism, innovation, and competition. With Comcast's demonstrated commitment to investment and innovation in communications, entertainment, and information, the new NBCU will be able to increase the quantity, quality, diversity, and local focus of its content, and accelerate the arrival of the multiplatform, "anytime, anywhere" future of video programming that Americans want. Given the intensely competitive markets in which Comcast and NBCU operate, as well as existing law and regulations, this essentially vertical transaction will benefit consumers and spur competition, and will not present any potential harm in any marketplace.

NBCU, currently majority-owned and controlled by GE, is an American icon – a media, entertainment, and communications company with a storied past and a promising future. At the heart of NBCU's content production is the National Broadcasting Company ("NBC"), the nation's first television broadcast network and home of one of the crown jewels of NBCU, NBC News. NBCU also has two highly regarded cable news networks, CNBC and

MSNBC. In addition, NBCU owns Telemundo, the nation's second-largest Spanish-language broadcast network, with substantial Spanish-language production facilities located in the U.S. NBCU's other assets include 26 local broadcast stations (10 NBC owned-and-operated stations ("O&Os"), 15 Telemundo O&Os, and one independent Spanish-language station), numerous national cable programming networks, a motion picture studio with a library of several thousand films, a TV production studio with a library of television series, and an international theme park business.

Comcast, a leading provider of cable television, high-speed Internet, digital voice, and other communications services to millions of customers, is a pioneer in enabling consumers to watch what they want, when they want, where they want, and on the devices they want. Comcast is primarily a distributor, offering its customers multiple delivery platforms for content and services. Although Comcast owns and produces some cable programming channels and online content, Comcast owns relatively few national cable networks, none of which is among the 30 most highly rated, and, even including its local and regional networks, Comcast accounts for a tiny percentage of the content industry. The majority of these content businesses will be contributed to the joint venture. The distribution side of Comcast (referred to as "Comcast Cable") is not being contributed to the new NBCU and will remain under Comcast's ownership and control.

The proposed transaction is primarily a *vertical* combination of NBCU's content with Comcast's multiple distribution platforms. Antitrust law, competition experts, and the FCC have long recognized that vertical combinations can produce significant benefits. They also have found that vertical combinations with limited horizontal overlaps generally do not threaten competition.

The transaction takes place against the backdrop of a communications and entertainment marketplace that is highly dynamic and competitive, and becoming more so every day. NBCU – today and post-transaction – faces competition from a large and growing roster of content providers. There are literally hundreds of national television networks and scores of regional networks. These cable networks compete for programming, for viewer attention, and for distribution on various video platforms, not only with each other but also with countless other video choices.

In addition, content producers increasingly have alternative outlets available to distribute their works, free from any purported "gatekeeping" networks or distributors. Today, NBCU has powerful marketplace incentives to purchase the best available programming, regardless of source. NBCU's programming schedule bears this out. Next week, third parties will own well over half of the 47 primetime (8-11pm) programs on NBC and its major cable channels (USA, Bravo, Oxygen, and SyFy). Post-transaction, the new NBCU will have the incentive and the financial resources to compete effectively with other leading content providers such as Disney/ABC, Time Warner, Viacom, and News Corp. by providing consumers the high-quality programming they want, and it will have no incentive – or ability – to restrict competition or otherwise harm the public interest.

Competition is fierce among distributors as well. Today, consumers in every geographic area have multiple choices of multichannel video programming distributors ("MVPDs") and can also obtain video content from many non-MVPDs. In addition to the local cable operator, consumers can choose from two MVPDs offering direct broadcast satellite ("DBS") service – DirecTV and Dish Network – which are now the second and third largest MVPDs in America, respectively. Verizon and AT&T, along with other wireline overbuilders, are strong, credible competitors, offering a fourth MVPD choice to tens of millions of American households and a fifth choice to some. Indeed, as competition among MVPDs has grown, Comcast's nationwide share of MVPD subscribers has steadily decreased (it is now less than 25 percent, a share that the FCC has repeatedly said is insufficient to allow an MVPD to engage in anticompetitive conduct). Moreover, current market dynamics are more telling than static measures of market shares; over the past two years, Comcast lost 1.2 million net video subscribers while its competitors continued to add subscribers – DirecTV, Dish Network, AT&T, and Verizon added 7.6 million net video customers over the same time period.

Consumers can also access high-quality video content from myriad other sources. Some households continue to receive their video through over-the-air broadcast signals, which have improved in quality and increased in quantity as a result of the broadcast digital television transition. Millions of households purchase or rent digital video discs ("DVDs") from one of thousands of national, regional, or local retail outlets, including Walmart, Blockbuster, and Hollywood Video, as well as Netflix, MovieCrazy, Café DVD, and others who provide DVDs by mail. High-quality video content also is increasingly available from a rapidly growing number of online sources that include Amazon, Apple TV, Blinkx, Blip.tv, Boxee, Clicker.com, Crackle, Eclectus, Hulu, iReel, iTunes, Netflix, Sezmi, SlashControl, Sling, Vevo, Vimeo, VUDU, Vuze, Xbox, YouTube – and many more. These sites offer consumers historically unprecedented quantities of professionally-produced content and user-generated content that can be accessed from a variety of devices, including computers, Internet-equipped televisions, videogame boxes, Blu-ray DVD players, and mobile devices. In addition, there is a huge supply of user-generated video content, including professional and quasi-professional content. YouTube, for example, which is by far the leader in the nascent online video distribution business, currently receives and stores virtually an entire day's worth of video content for its viewers *every minute*. And there are no significant barriers to entry to online video distribution. Thus, consumers have a staggering variety of sources of video content beyond Comcast and its rival MVPDs.

The video marketplace truly has no gatekeepers. As the United States Court of Appeals for the D.C. Circuit observed last year, "[T]he record is replete with evidence of ever increasing competition among video providers: Satellite and fiber optic video providers have entered the market and grown in market share since the Congress passed the 1992 [Cable] Act, and particularly in recent years. Cable operators, therefore, no longer have the bottleneck power over programming that concerned the Congress in 1992. Second, over the same period there has been a dramatic increase both in the number of cable networks and in the programming available to subscribers."

The combination of NBCU and Comcast's content assets under the new NBCU – coupled with management of the new NBCU by Comcast, an experienced, committed distribution innovator – will enable the creation of new pathways for delivery of content to consumers on a wide range of screens and platforms. The companies' limited shares in all relevant markets, fierce competition at all levels of the distribution chain, and ease of entry for cable and online programming ensure that the risk of competitive harm is insignificant. Moreover, the FCC's rules governing program access, program carriage, and retransmission consent provide further safeguards for consumers, as do the additional public interest commitments the companies have made to the FCC.

At the same time, the transaction's public interest benefits – particularly for the public interest goals of diversity, localism, competition, and innovation – are substantial. Through expanded access to outlets, increased investment in outlets, and lower costs, the new venture will be able to increase the amount, quality, variety, and availability of content, thus promoting *diversity*. This includes content of specific interest to diverse audiences, children and families, women, and other key audience segments. While NBCU and Comcast both already have solid records in creating and distributing diverse programming, the transaction will enable the new NBCU to expand the amount, quality, variety, and availability of content more than either company could do on its own. The new venture will also be able to provide more and better local programming, including local news and information programming, thereby advancing *localism*. The new NBCU and Comcast will be more innovative and effective players in video programming and distribution, spurring other content producers and distributors to improve their own services, thus enhancing *competition*. Marrying NBCU's programming assets with Comcast's multiple distribution platforms will make it easier for the combined entity to experiment with new business models that will better serve consumers, thus promoting *innovation*.

In addition, Comcast and NBCU have publicly affirmed their continuing commitment to free, over-the-air broadcasting. Despite a challenging business and technological environment, the proposed transaction has significant potential to invigorate NBCU's broadcasting business and expand the important public interest benefits it provides to consumers across this country. NBC, Telemundo, their local O&Os, and their local broadcast affiliates will benefit by having the full support of Comcast, a company that is focused entirely on entertainment, information, and communications and that has strong incentives – and the ability – to invest in and grow the broadcast businesses it is acquiring, in partnership with the local affiliates.

Moreover, combining Comcast's expertise in multiplatform content distribution with NBCU's extensive content creation capabilities and video libraries will not only result in the creation of more and better programming, but will also encourage investment and innovation, accelerating the arrival of the multiplatform, "anytime, anywhere" future of video programming that Americans want. This is because the proposed transaction will remove negotiation friction that currently inhibits the ability of Comcast to implement its pro-consumer vision of multiplatform access to quality video programming. Post-transaction, Comcast will have access to more content that it can make available on a wider range of platforms, including the new NBCU's national and regional networks and Comcast's cable systems and video-on-demand ("VOD") platform, and online. This increase in the value of services offered to consumers by the new company will stimulate competitors – including non-affiliated networks, nonaffiliated MVPDs, and the large and growing roster of participants in the video marketplace – to improve what they offer to consumers.

The past is prologue: Comcast sought for years to develop the VOD business, but it could not convince studio distributors – who were reluctant to permit their movies to be distributed on an emerging, unproven platform – to provide compelling content for VOD. This caution, though understandable in light of marketplace uncertainty, slowed the growth of an innovative and extremely consumer-friendly service. Comcast finally was able to overcome the contractual wrangling and other industry reluctance to participate in an innovative business model when it joined with Sony to acquire an ownership interest in Metro-Goldwyn-Mayer ("MGM"). This allowed Comcast to "break the ice" and obtain access to hundreds of studio movies that Comcast could offer for free on VOD. Thanks to Comcast's extensive efforts to foster the growth of this new technology, VOD has become very popular with consumers since it was invented in 2003 – the same year Apple unveiled the iTunes Music Store. Comcast customers have now used Comcast's VOD service more than 14 billion times – that's over 40 percent more than the number of downloads that consumers have made from the iTunes Store since 2003. By championing the growth of VOD, Comcast has been able to benefit not only its customers but also program producers, and it has stimulated other MVPDs to embrace the VOD model.

Similarly, there is every reason to believe that the transaction proposed here will create a pro-consumer impetus for making major motion pictures available sooner for in-home, on-demand viewing and for sustainable online video distribution – which, as the FCC has observed, will help to drive broadband adoption, another key congressional goal.

Comcast and the new NBCU will also be well positioned to help lead constructive efforts to develop consensus solutions to the problem of content piracy. NBCU has been a leading voice in the effort to reduce piracy in all its forms because it costs American jobs and trade opportunities. Comcast has consistently supported voluntary industry initiatives to deter piracy, educate consumers about copyright, and redirect them to legitimate sources of content. Together, the companies will redouble their efforts to persuade all the stakeholders to work together on the problem, while ensuring that consumer privacy and due process are always respected.

As noted above, the risk of competitive harm in this transaction is insignificant. Viewed from every angle, the transaction is pro-competitive:

First, combining Comcast's and NBCU's programming assets will give rise to no cognizable competitive harm. Even after the transaction, approximately six out of every seven channels carried by Comcast Cable will be unaffiliated with Comcast or the new NBCU. Comcast's national cable programming networks account for only about three percent of total national cable network advertising and affiliate revenues. While NBCU owns a larger number of networks, those assets account for only about nine percent of overall national cable network advertising and affiliate revenues. Therefore, in total, the new NBCU will account for only about 12 percent of total national cable network advertising and affiliate revenues. The new NBCU will rank as the fourth largest owner of national cable networks (measured by total revenues), behind Disney/ABC, Time Warner, and Viacom – which is *the same rank that NBCU has today*. Because both the cable programming market and the broader video programming market will remain highly competitive, the proposed transaction will not reduce competition or diversity, nor will it lead to higher programming prices to MVPDs, higher advertising prices to advertisers, or higher retail prices to consumers.

Second, Comcast's management and ownership interests in NBCU's broadcast properties raise no regulatory or competitive concern. While Comcast will own both cable systems and a stake in NBC owned-and-operated broadcast stations in a small number of Designated Market Areas ("DMAs"), the FCC's rules do not prohibit such cross-ownership, nor is there any policy rationale to disallow such relationships. Cross-ownership prohibitions that had been put in place decades ago have been repealed by actions of Congress, the courts, and the FCC. The case for any new prohibition, or any transaction-specific restriction, on cable/broadcast cross-ownership is even weaker today, given the increasingly competitive market for the distribution of video programming and robust competition in local advertising. And, importantly, each of the major DMAs in question has a significant number of media outlets, with at least seven non-NBCU over-the-air television stations in each DMA, as well as other media outlets, including radio. Thus, numerous diverse voices and a vibrantly competitive local advertising environment will remain following the combination of NBCU's broadcast stations and Comcast cable systems in each of the overlap DMAs.

Third, the combination of Comcast's and NBCU's Internet properties similarly poses no threat to competition. There is abundant and growing competition for online video content. The dominant leader in online viewing (by far) is Google (through YouTube and other sites it has built or acquired), with nearly 55 percent of online video viewing. This puts Google well ahead of Microsoft, Viacom, and Hulu (a service in which NBCU holds a 32 percent, non-controlling interest), and even farther ahead of Fancast (operated by Comcast, and currently at well below one percent). All of these services competing with Google have low- or mid-single digits shares of online video viewing. There are countless other sites that provide robust competition and near-infinite consumer choice. Even if one restricts the analysis to "professional" online video content, the combined entity will still have a small share and face many competitors. On the Internet, content providers essentially control their own destinies since there are many third-party portals as well as self-distribution options. Entry is easy. Thus, the transaction will not harm the marketplace for online video.

Finally, a vertical combination cannot have anticompetitive effects unless the combined company has substantial market power in the upstream (programming) or downstream (distribution) market, and such circumstances do not exist here. As noted, the video programming, video distribution, and Internet businesses are fiercely competitive, and the proposed transaction does not reduce that. competition. The recent history of technology demonstrates that distribution platforms are multiplying, diversifying, and increasingly rivalrous. Wired services have been challenged by both satellite and terrestrial wireless services. Cable has brought voice competition to the telephone companies; the telephone companies have added to the video competition that cable already faced; and both cable and phone companies are racing to deploy and improve broadband Internet. Static descriptions of markets have consistently failed to capture advances in distribution technologies. In this highly dynamic and increasingly competitive environment, speculative claims about theoretical problems arising from any particular combination should be subject to searching and skeptical scrutiny, given the accelerating power of technology to disrupt, continuously, all existing market structures.

In any event, there is a comprehensive regulatory structure already in place, comprising the FCC's program access, program carriage, and retransmission consent rules, as well as an established body of antitrust law that provides further safeguards against any conceivable vertical harms that might be presented by this transaction. The program access and program carriage rules address different aspects of the relationship between networks and MVPDs, and the retransmission consent rules address aspects of the relationship between MVPDs and broadcasters.

In a nutshell, the *program access* rules govern the process by which a satellite-delivered cable programming network that is affiliated with a cable operator sells its programming to MVPDs. These rules generally prohibit a cable operator from (i) unreasonably influencing whether an affiliated network sells its programming to an unaffiliated MVPD (or the terms on which it does so), (ii) unreasonably discriminating in the prices, terms, and conditions of carriage arrangements among competing MVPDs, and (iii) establishing exclusive contracts between satellite-delivered cable-affiliated programming networks and any cable operator.

The *program carriage* rules apply to the process by which a cable operator -- or any other MVPD -- buys cable programming from unaffiliated programmers. These rules generally prohibit MVPDs from (i) requiring an equity interest in a program network as a condition of carriage; (ii) coercing an unaffiliated program network to provide (or punishing an unaffiliated program network for not providing) exclusive rights as a condition of carriage; and (iii) unreasonably restraining the ability of an unaffiliated program network to compete fairly by discriminating on the basis of affiliation in the selection, terms, or conditions for carriage.

The *retransmission consent* rules generally require that broadcasters and MVPDs bargain in good faith over retransmission consent (i.e., the right to retransmit a broadcaster's signal). Like the program access rules, the good-faith bargaining rules generally ban exclusivity and unreasonable discrimination.

Although the competitive marketplace and regulatory safeguards protect against the risk of anticompetitive conduct, the companies have offered an unprecedented set of commitments to provide assurances that competition will remain vibrant. Comcast will commit voluntarily to extend the key components of the FCC's program access rules to negotiations with MVPDs for retransmission rights to the signals of NBC and Telemundo O&O broadcast stations for as long as the FCC's current program access rules remain in place (and Comcast has expressed a willingness to discuss with the FCC making the program access rules binding on it even if the rules were to be overturned by the courts).[1] Of particular note, Comcast will be prohibited in retransmission consent negotiations from unduly or improperly influencing the NBC and Telemundo stations' decisions about whether to sell their programming, or the terms and conditions of sale, to non-affiliated distributors. It would also shift to NBCU the burden of justifying any differential pricing between competing MVPDs. And the companies would accept the five-month "shot clock" that the Commission applies to program access adjudications that is intended to expedite resolution.

Moreover, the companies have offered concrete and verifiable commitments to ensure certain pro-consumer benefits of the transaction.

In addition to the commitment to continue to provide free, over-the-air broadcasting, mentioned previously, the companies have committed that following the transaction, the NBC O&O broadcast stations will maintain the same amount of local news and information programming they currently provide for three years following the closing of the transaction and will produce an additional 1,000 hours per year of local news and information programming for distribution on various platforms. The combined entity will maintain NBCU's tradition of independent news and public affairs programming and its commitment to promoting a diversity of viewpoints, maintaining the journalistic integrity and independence of NBCU's news operations.

The companies also have committed that, within 12 months of closing the transaction, Telemundo will launch a new Spanish language digital broadcast channel drawing on programming from Telemundo's library. Additionally, Comcast will use its On Demand and On Demand Online platforms to increase programming choices available to children and families, as well as to audiences for Spanish-language programming. Within three years of closing the transaction, Comcast has committed to add 1,500 additional programming choices appealing to children and families and 300 additional programming choices from Telemundo and mun2 to its VOD platforms. Comcast also will continue to provide free or at no additional charge the same number of VOD choices that it now provides, and will make available within three years of closing an additional 5,000 VOD choices over the course of each month that are available free or at no additional charge.

As Comcast makes rapid advances in video delivery technologies, more channel capacity will become available. So Comcast will commit that, once it has completed its digital migration company-wide (anticipated to be no later than 2011), it will add two new independently-owned and -operated channels to its digital line-up each year for the next three years on customary terms and conditions. Independent programmers would be defined as networks that (i) are not currently carried by Comcast Cable, and (ii) are unaffiliated with Comcast, NBCU, or any of the top 15 owners of cable networks, as measured by revenues.

With respect to public, educational, and governmental ("PEG") channels, Comcast has affirmatively committed not to migrate PEG channels to digital delivery on any Comcast cable system until the system has converted to all-digital distribution, or until a community otherwise agrees to digital PEG channels, whichever comes first. Comcast has also committed to innovate in the delivery of PEG content On Demand and On Demand Online.

We have proposed that these commitments be included in any FCC order approving the transaction and become binding on the parties upon completion of the transaction. A summary of the companies' commitments is attached to this testimony.

In the end, the proposed transaction simply transfers ownership and control of NBCU from GE, a company with a very diverse portfolio of interests, to Comcast, a company with an exclusive focus on, and a commitment to investing its resources in, its communications, entertainment, and information assets. This transfer of control, along with the contribution of Comcast's complementary content assets, will enable the new NBCU to better serve consumers. The new NBCU will advance key public policy goals: diversity, localism, competition, and innovation. Competition, which is already pervasive in every one of the businesses in which the new NBCU – and Comcast Cable – will operate, provides abundant assurance that consumer welfare not just be safeguarded, but increased. Comcast and NBCU will succeed by competing vigorously and fairly.

We intend to use the combined assets to accelerate and improve the range of choices that American consumers enjoy for entertainment, information, and communications. We would welcome your support.

COMCAST/NBCU TRANSACTION PUBLIC INTEREST COMMITMENTS

Comcast, GE, and NBC Universal take seriously their responsibilities as corporate citizens and share a commitment to operating the proposed venture in a way that serves the pubic interest. To demonstrate their commitment to consumers and to other media partners, the parties have made a set of specific, written commitments as part of their public interest filing with the Federal Communications Commission. Comcast, GE, and NBCU are committed to expanding consumer choice, ensuring the future of over-the-air broadcasting, enhancing programming opportunities, ensuring that today's highly competitive marketplace remains so, and maintaining journalistic independence for NBC's news properties. The parties' commitment to these principles will ensure that consumers are the ultimate beneficiaries of the proposed Comcast/NBCU transaction.

Applicants' Voluntary Public Interest Commitments

Local Programming

Commitment #1. The combined entity remains committed to continuing to provide free over-the-air television through its O&O broadcast stations and through local broadcast affiliates across the nation. As Comcast negotiates and renews agreements with its broadcast affiliates, Comcast will continue its cooperative dialogue with its affiliates toward a business model to sustain free over-the-air service that can be workable in the evolving economic and technological environment.

Commitment #2. Comcast intends to preserve and enrich the output of local news, local public affairs and other public interest programming on NBC O&O stations. Through the use of Comcast's On Demand and On Demand Online platforms, time slots on cable channels, and use of certain windows on the O&O schedules, Comcast believes it can expand the availability of all types of local and public interest programming.

- For three years following the closing of the transaction, NBC's O&O stations will maintain the same amount of local news and information programming that they currently provide.
- NBC's O&O stations collectively will produce an additional 1,000 hours a year of local news and information programming. This additional local content will be made available to consumers using a combination of distribution platforms.

Children's Programming

Commitment #3. Comcast will use its On Demand and On Demand Online platforms and a portion of the NBC O&Os' digital broadcast spectrum to speak to kids. Comcast intends to develop additional opportunities to feature children's content on all available platforms.

- Comcast will add 500 VOD programming choices appealing to children and families to its central VOD storage facilities within 12 months of closing and will add an additional 1,000 such VOD choices (for a total of 1,500 additional VOD choices) within three years of closing. (The majority of Comcast's cable systems will be connected to Comcast's central VOD storage facilities within 12 months of closing and substantially all will be connected within three years of closing.) Comcast will also make these additional choices available online to authenticated subscribers to the extent that Comcast has the requisite online rights.
- For three years following closing, each of NBC's O&O stations will provide one additional hour per week of children's educational and

informational programming utilizing one of the station's multicast channels.

Commitment #4. Comcast reaffirms its commitment to provide clear and understandable on-screen TV Ratings information for all covered programming across all networks (broadcast and cable) of the combined company, and to apply the cable industry's best-practice standards for providing on-screen ratings information in terms of size, frequency, and duration.

- NBCU will triple the time that program ratings remain on the air after each commercial break (from 5 seconds to 15 seconds).
- NBCU will make program ratings information more visible to viewers by using a larger format.

Commitment #5. In an effort to constantly improve the tools and information available for parents, Comcast will expand its growing partnership with Common Sense Media ("CSM"), a highly respected organization offering enhanced information to help guide family viewing decisions. Comcast will work to creatively incorporate CSM information it its emerging On Demand and On Demand Online platforms and other advanced platforms, and will look for more opportunities for CSM to work with NBCU.

- Comcast currently gives CSM content prominent placement on its VOD menus. Comcast and the new NBCU will work with CSM to carry across their distribution platforms more extensive programming information and parental tools as they are developed by CSM. Comcast and NBCU will explore cooperative efforts to develop . digital literacy and media education programs that will provide parents, teachers, and children with the tools and information to help them become smart, safe, and responsible users of broadband.
- Upon closing and pursuant to a plan to be developed with CSM, Comcast will devote millions of dollars in media distribution resources to support public awareness efforts over the next two years to further CSM's digital literacy campaign. The NBCU transaction will create the opportunity for CSM and Comcast to work with NBCU's broadcast networks, local broadcast stations, and cable networks to provide a targeted and effective public education campaign on digital literacy, targeting underserved areas, those with high concentrations of low-income residents and communities of

color, as well as target Latino communities with specifically tailored Spanish-language materials.

Programming for Diverse Audiences

Commitment #6. Comcast intends to expand the availability of over-the-air programming to the Hispanic community utilizing a portion of the digital broadcast spectrum of Telemundo's O&Os (as well as offering it to Telemundo affiliates) to enhance the current programming of Telemundo and mun2.

- Within 12 months of closing the transaction, Telemundo will launch a new Spanish language channel using programming from Telemundo's library that has had limited exposure, to be broadcast by each of the Telemundo O&O stations on one of their multicast channels. The Telemundo network also will make this new channel available to its affiliated broadcast stations on reasonable commercial terms.

Commitment #7. Comcast will use its On Demand and On Demand Online platforms to feature Telemundo programming.

Commitment #8. Comcast intends to continue expanding the availability of mun2 on the Comcast Cable, On Demand, and On Demand Online platforms.

- Comcast will increase the number of VOD choices from Telemundo and mun2 available on its central VOD storage facilities from approximately 35 today, first to 100 choices within 12 months of closing and then to a total of 300 additional choices within three years of closing. Comcast will also make these additional choices available online to its subscribers to the extent that it has the requisite online rights.

Expanded Video On Demand Offerings at No Additional Charge

Commitment #9. Comcast currently provides approximately 15,000 VOD programming choices free or at no additional charge over the course of a month. Comcast commits that it will continue to provide at least that number of VOD choices free or at no additional charge. In addition, within three years of closing the proposed transaction, Comcast will make available over the course of a month an additional 5,000 VOD choices via its central VOD storage facilities for free or at no additional charge.

Commitment #10. NBCU broadcast content of the kind previously made available at a per-episode charge on Comcast's On Demand service and currently made available at no additional charge to the consumer will continue to be made available at no additional charge for the three-year period after closing.

Public, Educational, and Governmental ("PEG") Channels

Commitment #11. With respect to PEG channels, Comcast will not migrate PEG channels to digital delivery on any Comcast cable system until the system has converted to all-digital distribution (i.e. until all analog channels have been eliminated), or until a community otherwise agrees to digital PEG channels, whichever comes first.

Commitment #12. To enhance localism and strengthen educational and governmental access programming, Comcast will also develop a platform to host PEG content On Demand and On Demand Online within three years of closing.

- Comcast will select five locations in its service area to test various approaches to placing PEG content on VOD and online. Comcast will select these locations to ensure geographic, economic and ethnic diversity, with a mix of rural and urban communities, and will consult with community leaders to determine which programming – public, educational and/or governmental – would most benefit local residents by being placed on VOD and online.
- Comcast will file annual reports to inform the Commission of progress on the trial and implementation of this initiative.

Carriage for Independent Programmers

Commitment #13. As Comcast makes rapid advances in video delivery technologies, more channel capacity will become available. So Comcast will commit that, once it has completed its digital migration company-wide (anticipated to be no later than 2011), it will add two new independently-owned and -operated channels to its digital line-up each year for the next three years on customary terms and conditions.

- New channels are channels not currently carried on any Comcast Cable system.
- Independent programmers are entities that are not affiliated with Comcast, NBCU, or any of the top 15 owners of cable networks (measured by revenue).

Expanded Application of the Program Access Rule Protections

Commitment #14. Comcast will commit to voluntarily accept the application of program access rules to the high definition (HD) feeds of any network whose standard definition (SD) feed is subject to the program access rules for as long as the Commission's current program access rules remain in place.

Commitment #15. Comcast will commit to voluntarily extend the key components of the FCC's program access rules to negotiations with MVPDs for retransmission rights to the signals of NBC and Telemundo O&O stations for as long as the Commission's current program access rules remain in place.

- Comcast will be prohibited in retransmission consent negotiations from unduly or improperly influencing the NBC and Telemundo O&O stations' decisions about the price or other terms and conditions on which the stations make their programming available to unaffiliated MVPDs.
- The "burden shifting" approach to proof of discriminatory pricing in the program access rules will be applied to complaints regarding retransmission consent negotiations involving the NBC and Telemundo O&O stations.
- The five-month "shot clock" applied to program access adjudications would apply to retransmission consent negotiations involving the NBC and Telemundo O&O stations.

Journalistic Independence

Commitment #16. The combined entity will continue the policy of journalistic independence with respect to the news programming organizations of all NBCU networks and stations, and will extend these policies to the potential influence of each of the owners. To ensure such independence, the combined entity will continue in effect the position and authority of the NBC News ombudsman to address any issues that may arise.

Labor-Management Relations

Commitment # 17. Comcast respects NBCU's existing labor-management relationships and expects them to continue following the closing of the transaction. Comcast plans to honor all of NBCU's collective bargaining agreements.

End Notes

[1] In October 2007, the FCC released an Order extending for an additional five years the ban on exclusive contracts between vertically integrated programmers and cable operators -- the one portion of the program access rules that Congress had slated to sunset in 2002. On appeal, Cablevision and Comcast have argued that the FCC applied an incorrect standard governing the circumstances under which the FCC may prevent the exclusivity rule from sunsetting automatically; and that the FCC was required to let the rule sunset, or at least narrow it. Comcast was motivated in large part by the inequity of applying an anti-exclusivity rule to cable, while our satellite competitors are able to use exclusive programming contracts against us. Oral argument was held on September 22, 2009. Contrary to the claims of some outside parties, Comcast has not challenged all of the features of the program access rules in this litigation or asserted that the exclusivity ban, or any other portion of the program access rules, is unconstitutional. Rather, we have challenged only the extension of the exclusivity ban, and have reminded the FCC and the courts that they must take the First Amendment into account when they make, review, or apply the program access rules.

In: Media Industry Programming, Competition...   ISBN: 978-1-61122-078-0
Editors: Ryan E. Moore   ©2011 Nova Science Publishers, Inc.

Chapter 5

TESTIMONY OF THOMAS W. HAZLETT, PANEL ON THE COMCAST-NBCU VENTURE, BEFORE THE COMMITTEE TO THE JUDICIARY, HEARING ON "COMPETITION IN THE MEDIA AND ENTERTAINMENT DISTRIBUTION MARKET"

Thomas W. Hazlett

I. INTRODUCTION

My name is Thomas W. Hazlett. I am a professor of law & economics at George Mason University, where I head the Information Economy Project. I formerly served as Chief Economist of the Federal Communications Commission, and am a columnist for the *Financial Times*. I have written widely on the economics of telecommunications markets and the effect of government regulation in the sector. I am also the author of PUBLIC POLICY TOWARD CABLE TELEVISION, with Matthew L. Spitzer (MIT Press; 1997).

In the proposed transaction being discussed here today, Comcast becomes a 51% owner of Newco, with General Electric receiving 49%. The new venture will combine Comcast's cable TV program networks with NBCU's broadcasting network, cable networks and broadcast TV stations, along with Telemundo's broadcast network and stations. In addition, other assets of NBC, including the Universal Studios theme park, will be contributed to the enterprise.

The transaction is large, but not among the largest mergers historically. The joint venture is estimated to be worth about $28 billion, less than the $34 billion Viacom purchase of CBS in 1999, for example, or the $35 billion Sprint purchase of Nextel in 2005. And the 51% Comcast stake, at about $15 billion, is much less than these and many other corporate transactions.

II. COMPETITIVE ANALYSIS

The economic policy question concerns how the deal impacts market competition. On that score, the issues are straightforward. The merger is primarily a vertical combination where Comcast, a cable operator distributing video programming to millions of household subscribers, is acquiring ownership of additional programming assets. This does not lessen competition in any market, but allows the content distributor to achieve efficiencies by producing complementary products.

There are special cases in which vertical integration can lead to anti-competitive foreclosure, but the evidence indicates that these special circumstances do not apply. Studies of vertical integration in cable generally confirm the baseline analysis: efficiencies typically result when firms elect to combine programming and distribution.[1] As an empirical matter, the trend in the sector is away from vertical integration, meaning that operators do not believe that they can increase profits via vertical foreclosure. The ownership of cable program networks has sharply declined over the past two decades; the spin-off of cable TV systems by Viacom (in 1996) and Time Warner (in 2008) are key components of this trend.

In video programming, there is a horizontal aspect to the combination: Comcast currently owns some cable network assets, and those will merge with direct rivals owned currently by General Electric. But the Comcast share is meek; combined with NBC-Universal program assets it will account for only about 12% of total U.S. cable program network revenues. This will yield some economies of scale, or so Comcast hopes, but it hardly moves the needle in terms of the concentration of the industry. The GE-owned cable assets are smaller, in total, than those owned by Disney, Time Warner, and Viacom, and will – with Comcast's assets – remain so.

The very good news for consumers (and programmers) in recent years is that local market competition has taken off. Twenty years ago, one local cable TV system dominated multi-channel video program distribution in each franchise area. Today, there are about 3.4 competitors per market: the local cable operator, two satellite TV rivals (each with a national footprint), and – in nearly half the country – a telco TV provider.

Even before thinking about the next generation of broadband video, the market power of incumbent cable systems has been dealt a lethal blow. This is seen in examining market prices for cable TV systems, now selling (in constant dollars) for about what they sold for in 1990. See Figure 1. This is despite the fact that systems now deliver not just video but broadband data and voice, the "triple play," and that they deliver hundreds more channels to subscribers. These modern, two-way, high-capacity digital platforms are substantially more costly to build, meaning that the returns realized by cable system investors are a fraction of what they were a generation ago.[2] This is directly attributable to the outbreak of competitive rivalry. Nothing in the Comcast-GE deal threatens to disturb that trend.

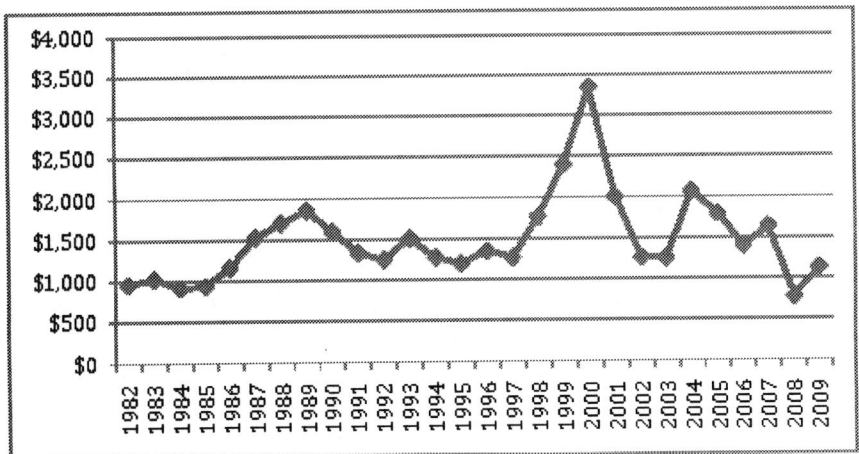

Source: CABLE TV INVESTOR, SNL Kagan (Jan. 25, 2010).

Figure 1. Value per cable tv subscriber (constant 1982-84 dollars)

III. A QUESTION OF BUSINESS STRATEGY

In acquiring additional programming assets, Comcast swims against the tide. The company is wagering that it can make more productive use of GE's cable and broadcast networks. It does so knowing that its markets are in tumult. Video products are jumping from platform to platform – not just from cable to satellite, but from television to broadband, from linear channels to on-demand networks, from pay to freemium, from TV screens to mobile devices. Some financial analysts have praised Comcast for its bold new enterprise; many have condemned it. *Didn't they learn anything from the failed AOLTime Warner merger?* is a fairly popular reaction.

The simple fact is that no one fully understands where today's tide is headed. Cable operators do not know if they need fear Verizon or Echostar, Google or Apple. Time Warner believes that splitting its cable operations from its program ownership is the best way to prepare for the coming storm. Comcast has come to a much different conclusion. Markets allow these rival strategies to be tested and winning strategies rewarded. I wish Comcast and General Electric shareholders well in their educated guesses.

End Notes

[1] Tasneem Chipty, *Vertical Integration, Market Foreclosure, and Consumer Welfare in the Cable Television Industry*, 91 AMERICAN ECONOMIC REVIEW 428 (Jun. 2001); George S. Ford and John D. Jackson, *Horizontal Concentration and Vertical Integration in the Cable Television Industry*, 12 REVIEW OF INDUSTRIAL ORGANIZATION 501 (Aug. 1997); Austan Goolsbee, *Vertical Integration and the Market for Broadcast and Cable Television Programming*, paper for the Federal Communications Commission (Sept. 5, 2007); Thomas W. Hazlett, *Vertical Integration in Cable Television: The FCC Evidence*, paper submitted to the FCC (Oct. 19, 2007).

[2] Thomas W. Hazlett & Dennis L. Weisman, *Market Power in U.S. Broadband Services*, George Mason University Law and Economics Research Paper Series 09-69 (Nov. 2009); http://mason.gmu.edu/~thazlett/pubs/Hazlett.Weisman.Broadband.SSRN-id152556 8.pdf.

In: Media Industry Programming, Competition... ISBN: 978-1-61122-078-0
Editors: Ryan E. Moore ©2011 Nova Science Publishers, Inc.

Chapter 6

TESTIMONY OF DR. MARK COOPER, DIRECTOR OF RESEARCH, CONSUMER FEDERATION OF AMERICA, BEFORE THE COMMITTEE TO THE JUDICIARY, HEARING ON "COMPETITION IN THE MEDIA AND ENTERTAINMENT DISTRIBUTION MARKET"

Mark Cooper

Mr. Chairman and Members of the Committee,

My name is Dr. Mark Cooper. I am the Director of Research at the Consumer Federation of America. I appear before you today on behalf of the Consumer Federation of America, Free Press and Consumers Union. We appreciate the opportunity to share our views on media markets and a merger that is unique in the history of the video market, one that will go a long way toward determining whether or not the future of video viewing in America is more competitive and consumer-friendly than the past.

The merger of Comcast and the National Broadcasting Company (NBC) is a hugely complex undertaking, unlike any other in the history of the video marketplace. Allowing the largest cable operator in history to acquire one of the nation's premier video content producers will radically alter the structure of the video marketplace and result in higher prices and fewer choices for consumers. The merging parties are already among the dominant players in the current video market. This merger will give them the incentive and ability to not only preserve and exploit the worst aspects of the current market, but to extend them to the future market.

Comcast has sought to downplay the impact of the merger by claiming that it is a small player in comparison to the vast video universe in which it exists. It has also glossed-over the fact that this merger involves the elimination of actual head-to-head competition. Finally, it has argued that existing protections and public interest promises will prevent any harms that might result from the merger. All three claims are wrong.

Neither Comcast's regurgitation of market shares and counts of outlets and products, nor its public interest commitments begin to address the fundamental public policy questions and competitive issues at stake in this merger. Nor can the merger of these companies be viewed separately from the products they sell. NBC and Comcast do not sell widgets. They sell news and information and access to the primary platforms American use to receive this news and information. Control over production and distribution of information has critical implications for society and democracy. As a consequence, the merger of these two media giants reaches far beyond the economic size of the merging parties to the very content consumers receive, and how they are permitted to access it.

Finally, if the size and scope of this merger is not sufficient to give you pause, the past actions of the acquiring party should. Comcast has raised cable rates for consumers every year, and is among the lowest ranked companies in terms of customer service. Comcast is the frequent subject of program access complaints of competing video providers, as well as of discriminatory carriage complaints by independent programmers. Finally, Comcast is on record lying to a federal agency regarding whether they blocked Internet users' access to a competing a video application for anti-competitive purposes. These past practices do not bode well for future competition if Comcast is allowed to acquire NBC. Further, Comcast's lack of candor in past proceedings cast doubt on the prudence of relying on Comcast's voluntary public interest commitments as a means of addressing the anti-consumer impacts of this merger.

The goal of mega-mergers such as this is to cut costs and increase revenues. The most direct path to those outcomes are firing workers and raising prices. Cutting jobs is hardly a laudable goal in the current environment, but the primary "synergy" that mergers produce is the ability to reduce employment by sharing resources between the commonly-held companies. To expect the opposite to happen here based on the evidence-free assertions of Comcast would be foolhardy. Simply put, this merger is about higher prices, fewer choices, and lost jobs.

THE BIGGEST GETS BIGGER (AND STRONGER)

Comcast is the nation's largest cable operator, largest broadband service provider and one of the leading providers of regional cable sports and news networks. NBC is one of only four major national broadcast networks, the third largest major owner of local TV stations in terms of audience reach, an icon of local and national news production and the owner of one of a handful of major movies studios.

As large as Comcast is nationally, it is even more important as a local provider of video services. Comcast is a huge entity in specific product markets. It is the dominant multi-channel video programming distributor (MVPD) in those areas where it holds a cable franchise, accounting, on average for over half of the MVPD market. It is the dominant broadband access provider in the areas where it has a cable franchise, accounting for over half of that market. This dominance of local market distribution platforms is the source of its market power. The merger will eliminate competing distribution platforms in some of its markets and will give Comcast control over strategic assets to preserve and expand its market power in all of its markets.

Broadcasters and cable operators are producers of goods and services that compete head-to-head, including local news, sports, and advertising. In addition, NBC and Comcast are also suppliers of content and distribution platforms, which are goods and services that complement one another. In both roles there is a clear competitive rivalry between them. For example, in providing complementary services, broadcasters and cable operators argue about the price, channel location and carriage of content. The merger will eliminate this natural rivalry between two of the most important players in the multi-channel video space, a space in which there are only a handful of large players.

These anticompetitive effects of the merger are primarily what antitrust practice refers to as horizontal effects, as shown in Exhibit 1. They are likely to reduce competition in specific local markets — head-to-head competition in local video markets, head-to-head competition for programming viewers, head-to-head competition for distributions platforms. The merger will raise barriers to entry even higher through denial and manipulation of access to programming and the need to engage in two-stage entry. The merger will increase the likelihood of the exercise of existing market power within specific markets, and will increase the incentive and ability to raise prices or profits.

The fact that some of the leverage is brought to bear because of the link to complementary products (i.e. is vertical in antitrust terms), should not obscure the reality that the ultimate effects are on horizontal competition in both the distribution and programming markets. The merger would dramatically increase the incentive and ability of Comcast to raise prices, discriminate in carriage, foreclose and block competitive entry and force bundles on other cable systems. The merger enhances the ability of Comcast to preserve its position as the dominant local MVPD, reinforce its ability to exercise market power in specific cable or programming markets and extend its business model to the Internet.

We raise these concerns about the merger based on eight specific anti-competitive effects that the merger will have on the video market. The attached exhibit presents the list of distribution and content assets owned in whole or in part by these two companies. The exhibit makes it crystal clear that they do compete head-to-head across a number of product and geographic markets and the assets represent an arsenal of complements that would be powerful ammunition to use as leverage against existing competitors and new entrants.

HIGHER PRICES, FEWER CHOICES, LESS COMPETITION

(1) **This Merger will reduce choice and competition in local markets.**
The merging parties currently compete head-to-head as distributors of video content, in local markets. Because broadcasters own TV stations, they compete with cable in local markets for audiences and advertisers — especially in the production and distribution of local news, and local and political advertising. This merger eliminates this head-to-head competition in 11 major markets where NBC owns broadcast stations and Comcast operates a cable franchise. These 11 markets account for nearly a quarter of U.S. TV households.

This merger also eliminates a competitor for local and political advertising. In fact, in 2006 NBC told the Federal Communications Commission that local cable operators present the single biggest threat to broadcasters in terms of securing local and political advertising. The concentration of local markets and increase in concentration created by this merger, as measured by local advertising vastly exceed the level that should trigger close antitrust scrutiny under the DOJ/FTC *Merger Guidelines*.[1] Now that NBC is looking to merge with Comcast, the potential elimination of this local competition has been conveniently ignored. But federal authorities cannot and should not ignore the fact that a merger between Comcast and NBC is likely to cause a significant decline in competition in local advertising markets and excessive domination by the merged company. Not only will advertisers lose an important option, but also the merger will be to the detriment of other local broadcasters - particularly smaller, independent ones - who are already facing ad revenue declines in an economic downturn. A stand-alone broadcaster will not be able to offer package deals and volume discounts for advertising across multiple channels the way that Comcast/NBC will be able to do post-merger. That means other local broadcasters will have less money to produce local news and hire staff. To compete, rival broadcasters will have two options: fire staff and reduce production of local news and information; or consolidate in order to compensate for market share lost to the new media mammoth.

(2) **This merger removes an independent outlet and an independent source of news and information.** These two companies compete in the video programming market, where Comcast's regional sports and news production compete with NBC's local news and sports production. By acquiring NBC, Comcast's incentive to develop new programming would be reduced. Instead of continuing to compete to win audience, it just buys NBC's viewers. Where two important entities were producing programming, there will now be one.

(3) **The merger will eliminate competition between Comcast and NBC in cyberspace.** NBC content is available online in a variety of forms and on different websites and services. Most prominently, of course, NBC is a stakeholder in Hulu — an online video distribution portal that draws millions of viewers. Comcast has put resources into developing its own online video site - "Fancast" — where consumers

can find content owned by the cable operator. The merger eliminates this nascent, head-to-head competition.

Moreover, Comcast is the driving force behind the new "TV Everywhere" initiative. This collusive venture — which we believe merits its own antitrust investigation -- would tie online video distribution of cable content to a cable subscription and pressure content providers to restrict or refrain from online distribution outside of the portal. This is a disaster for video competition. The proposed merger strengthens Comcast's hand in this scheme by increasing their market power in both traditional and online video distribution. Comcast is clearly attempting to control the distribution of the video content it makes available on the web by restricting sales exclusively to Comcast cable customers. It does not sell that content to non-Comcast customers. By contrast, NBC has exactly the opposite philosophy -- or at least it did. Through Hulu, NBC is competing for both Comcast and non-Comcast customers by selling video online that is not tied to cable. NBC also has incentives to make its programming available in as many points of sale as possible. Merger with Comcast will put an end that pro-competitive practice. "TV Everywhere" is a blatant market division scheme intended to extend the cable "non-compete" regimen from physical space to cyberspace.

(4) The merger will provide Comcast with greater means to deny rivals access to Comcast controlled programming. Comcast already has incentive to undermine competing cable and satellite TV distributors by denying them access to critical, non-substitutable programming, or by extracting higher prices from competitors to induce subscribers to switch to Comcast. Post-merger it will have a great deal more content to use as an anticompetitive tool. Comcast has engaged in these anticompetitive acts in the past and by becoming a major programmer it will have a much larger tool to wield against potential competitors. Moreover, Comcast has opposed, and is currently challenging in court, the few rules in place that would prevent it from withholding its programming from competing services. Strangely enough, Comcast's CEO promised members of Congress in a previous hearing that the company would continue to abide by these rules even if they were successful in getting the court to throw them out. Yet Comcast continues to spend shareholder dollars trying to overturn an FCC regulation that it promises to follow regardless of the case's outcome. As a show of good faith, we have asked Comcast to withdraw its suit. In response Comcast has equivocated. Now it claims it made no such promise.

(5) **The merger will provide greater incentive for Comcast to discriminate against competing independent programmers.** Comcast already has a strong incentive to, and significant track record of, favoring its own programming over the content produced by others with preferential carriage deals. Post-merger it will have a lot more content to favor. The current regulatory structure does not appear sufficient to remedy the existing problem and cannot be expected to address the resulting post-merger threat to independent programmers. The econometric analysis of program carriage indicates there is a great deal of discrimination occurring already. The fact that the FCC is continually trying to catch up with complaints of program carriage discrimination is testimony to the existence of the problem and the inability of the existing rules to correct it.

(6) **The merger will stimulate a domino effect of concentration between distributors and programmers.** The new combination will create a major asymmetry in the current cartel model in the cable industry. It brings together a large cable provider with a huge stable of must-have programming *and* the largest wireline broadband platform in America. Very likely, this will trigger more mergers and acquisitions because it changes the dynamics of the market. But there will be no positive competitive outcomes resulting from this change. This merger signals that the old, anticompetitive game is still on -- but with a twist. Like all other cable operators, Comcast has never entered the service territory of a competing multichannel video program provider, allowing everyone to preserve market power and relentlessly raise prices. But Comcast's expanded assets and especially its new leverage over the online video market will give it a substantial edge against its direct competitors in its service territory. The likely effect of the merger will be for other cable distribution and broadband companies to muscle up with their own content holdings to try and offset Comcast's huge advantage. In other words, there is only one way to deal with a vertically integrated giant that has must-have content and control over two distribution platforms — you have to vertically integrate yourself. This merger would send a signal to the industry that the decades old game of mutual forbearance from competition will be repeated but at the next level of vertical integration that spills over into the online market. Watch for AT&T

and Verizon to be next in line for major content acquisitions. When that happens, it will be extremely difficult for any company that is merely a programmer or merely a distributor to get into the market. Barriers to entry to challenge vertically integrated incumbents will be nearly unassailable. The only option may be a two-stage entry into both markets at the same time — which is an errand reserved only for the brave and the foolish.

(7) **By undermining competition this merger will result in higher prices for consumers.** Comcast already raises its rates every year for its cable subscribers, and prices are likely to rise further after the merger. By weakening competition, Comcast's market power over price is strengthened, but there are also direct ways the merger will push the price to consumers up. Comcast will have the opportunity and incentive to charge its competitors more for NBC programs and force competitors to pay for less desirable Comcast cable channels in order to get NBC programming -- those added costs will mean bigger bills for cable subscribers. Furthermore, the lack of competitive pressure that has failed to produce any appreciable downward pressure on cable rates since 1983, will not discipline Comcast from raising its own rates.

(8) **This merger will result in higher prices for consumers through the leveraging of "retransmission rights."** Through its takeover of local NBC broadcast stations, Comcast will also gain special "retransmission consent rights," which allow stations to negotiate fees for cable carriage of broadcast signals. These rights will enable Comcast to leverage control over must-have local programming and larger bundles of cable channels to charge competing cable, telco and satellite TV providers more money for content. Additionally, once Comcast acquires a broadcaster, it will have the means and incentive to raise retransmission rights payments for NBC-owned stations. This will be reinforced by two factors. First, as the owner of NBC, Comcast profits from the retransmission payments it receives and does not lose from the retransmission payments it makes, which are passed through to consumers. Second, Comcast can charge competitors more for local NBC programming, and will be able to exploit asymmetric information. Cable operators do not publish what they pay for retransmission; broadcasters do not publish what they get. Because of Comcast's superior bargaining power, it will ask for more and pay less.

EMPIRICALLY GROUNDED, RESPONSIBLE MERGER ANALYSIS V. "DO NOTHING THEORY"

In response to my February 4, 2010 testimony in the House Commerce Committee and the Senate Judiciary Committees, the Free State Foundation has posted a rebuttal by Richard Epstein, a law professor at the University of Chicago and a Senior Fellow at the Hoover Institution.[2] His response to my testimony is an example of the predictable chorus of free market ideologues who inevitably parrot the claims of the merging parties that new efficiencies will benefit consumers and that there is more than enough competition to prevent abuses.

Thankfully, the era of "don't worry, be happy" antitrust enforcement in America is over.[3] Professor Epstein's approach to merger analysis reflects all of the worst weaknesses of the Chicago School approach that he espouses. It is based on pure theory, no facts.[4] Moreover, it is premised on a theory that is biased toward the approval of mergers[5] because it favors the creation of monopoly rents[6] by dominant firms[7] and ignores the importance of dynamic efficiency and disruptive entrants and mavericks.[8]

Professor Epstein ignores the mountain of evidence that there are numerous clearly defined markets in which Comcast and NBC compete head-to-head. In part this stems from the fact that he never attempts to define product and geographic markets. This failure is rooted conceptual and empirical flaws in his approach. On the one hand, the Chicago School approach assumes that self-correcting markets will automatically respond to the market power created by mergers,[9] because entry is easy.[10] One the other hand, the approach defines markets too broadly[11] and underestimates the importance of horizontal market power.[12]

Efficiency gains and benefits are overblown in the Chicago School approach. Indeed, they are used as an excuse to justify market power, rather than an empirically demonstrated fact.[13] All merging parties claim efficiency gains and "synergies", though few actually deliver on those promises. Nevertheless, the Chicago School treats those claims as a bona fide magic wand that blesses every merger that comes along.[14] Professor Epstein provides no evidence of efficiency gains or that the assumed benefits will be passed on to consumers and ignores the importance of wealth transfers as a consumer harm that can result from mergers, weaknesses that are endemic to this school of thought.[15]

The theoretically induced blindness to horizontal problems of this merger is matched by the utter ignorance of the vertical problems that it poses.[16] Abuse of vertical leverage has long been recognized as a critical problem that is ignored by Chicago School theory. [17] The cable industry has long been afflicted by the use of vertical leverage to undermine horizontal competition and Comcast has been in the forefront of that practice.[18] Empirical studies have repeatedly shown that by discriminating against independent programmers in affording carriage, cable operators have advanced the interest of their own programming and undermined the prospect for independent programming, impairing competition in content markets. By denying competing distribution platforms access to video content, cable operator have retarded competition in the distribution market, a practice that has led to repeated disputes at the Federal Communications Commission.

The bitter fruit of lax, "don't worry, be happy" antitrust enforcement has been tasted by the public in the approval of a string of mergers that have allowed the MVPD market to become concentrated and sustained the constant increase in prices in the cable industry. Professor Epstein asks us to ignore this central fact of life in the MVPD market because Chicago School Theory pays little attention to consumer welfare.[19] Responsible antitrust authorities cannot do so.

The track record of past mergers and merger conditions has become a bone of contention in the Comcast NBC case. In a thin attempt to soothe worries regarding the merger, merger supporters have listed a number of recent media and communications mergers, which they claim, did not result in the sky falling-in on consumers (to wit, AT&T-SBC, Verizon-MCI, News Corp.- DirecTV, AOL-Time Warner, XM-Sirius). However, in referencing past mergers a defense, supports of the present merger draw the wrong conclusions in four crucial respects.

First, these mergers pale in comparison to consolidation of control over both programming production and distribution that would occur as a result of a Comcast takeover of NBC. The Comcast-NBC merger is much larger and involves uniquely anticompetitive threats resulting from the marriage of a major video content producer to the nation's largest cable television provider and broadband service provider.

Second, many of these past mergers were prevented from doing their worst because, in every case, antitrust authorities imposed important conditions to prevent the anticompetitive, anti- consumer harms that the consolidation would have produced. These conditions were, of course, opposed by the Chicago School ideologues, just as they now oppose the imposition of any conditions on the current merger.

Third, virtually all of these mergers all resulted in consumer harm, even in spite of conditions that helped to mitigate the damage to some extent. The telecom mergers, in particular were disastrous for consumers. They eliminated major competitors in the marketplace for wireline broadband service, reversed the outcomes of the pro-competitive breakup of AT&T and the pro-competitive 1996 Telecommunications Act, and delivered a wireline duopoly that has resisted meaningful price competition ever since. These mergers also resulted in massive consolidation in the wireless industry (by virtue of granting huge market power to these wireline companies that also had wireless services) — pushing AT&T and Verizon into dominant positions that are quickly giving us the same problems in mobile communications.

Finally, these mergers did not produce the synergies and efficiencies that these companies promised. Instead, the claims of efficiency, that were used to justify mergers in the past decade, were vastly overblown or failed to materialize at all. The "efficient market hypothesis" at the center of the Chicago School analytic framework, which allowed companies to wave a magic efficiency wand and blind the antitrust authorities to the anticompetitive impact of merger, was the cornerstone of the "don't' worry, be happy" era. The "efficient market hypothesis" is crumbling; buried, if not dead, beneath the rubble of the financial system.[20]

A COMCAST/NBC MERGER SHOULD NOT BE ALLOWED TO PROCEED WITHOUT MAJOR STRUCTURAL REFORMS OF THE VIDEO MARKET

The merger has so many anti-competitive, anti-consumer, and anti-social effects that it cannot be fixed. Comcast's claim that FCC oversight will protect the public is absurd. Moreover, such claims are undercut by the fact that Comcast is presently opposing the very rules it says will prevent it from anticompetitive conduct. The challenges that this merger poses to the future of video competition cannot be ignored, or brushed aside by reliance on FCC rules that have yet to remedy current problems and, thus, are ill-equipped to attend to the increased anticompetitive means and incentives that will result from Comcast's acquisition of NBC. The FCC rules have failed to break the stranglehold of cable to-date; there is no reason to believe they will be better able to tame the video giant that will result from this merger.

Further, any suggestion that the public interest commitments Comcast has made will solve these problems is misguided. Temporary band-aids cannot cure long-term structural injuries. Comcast's promises lack substance and accountability. More importantly, the commitments do not begin to address the anticompetitive effects of the merger. Many of Comcast's commitments amount to little more that a promise to obey the law. Where they go beyond current law, they largely fall within the company's existing business plans. Anything beyond that is meager at best, and in no way substitutes for the localism and diversity that a vigorously competitive industry would produce.

We recognize that the company has made some promises that address some specific concerns of members of the Congress and this committee. We appreciate the fact that everyone recognizes that those special interest promises are far from adequate to protect the interests of the broader public. So in my remarks today I will take up the challenge that some members of the Committee have laid down in terms of identifying the conditions that would begin to address the broader problems with this merger and in this industry. I emphasize the structure and process of enforcement of conditions, rather than the details.

First, all of the major areas of competitive concern should be addressed, in addition to the localism and diversity areas that Comcast has admitted are a problem — local markets/affiliate relations, cable program access, cable carriage, Internet distribution, independent programming in broadcast and prime time. If federal authorities allow this merger to go forward, they should not merely impose conditions on the merger, they should reform the regulatory structure of the industry to address the underlying problems that this merger will make much worse. The only way to address the harm that this merger will do to competition and consumers is to address the underlying problems that afflict video consumers in America.

To ensure that the conditions are enforceable, we believe that the federal authorities with oversight over this merger should complete industry-wide proceedings that address the underlying problems before the merger is approved. In every one of the areas where we believe that broad public interest is at risk, there is a pending proceeding or complaint that provides the opportunity to quickly and effectively address the underlying problems in the industry that would be made so much worse by this merger. When it comes to relations between the networks and their affiliates, cable program access, cable program carriage, and independent programming on broadcast networks, the

FCC has available vehicles to move quickly to adopt strong rules to protect the public. The antitrust authorities have been asked to examine the anti-consumer, anticompetitive market division scheme Comcast is pushing for Internet distribution of video content. These agencies should act to outline the rules of the road and create the institutional structures that will prevent the abuse of market power and promote competition in the MVPD market.

Once these industry-wide mechanisms are in place, the agencies should then consider whether additional conditions are necessary to meet the unique threat to competition and the public interest embodied in this merger.

Finally, federal authorities must not only impose meaningful conditions with enforceable sanctions, but the Comcast should also agree not to challenge the legality of the conditions or render aid and comfort to those who do. If they challenge the legality of the regulatory mechanisms that underlie any of the major conditions imposed on the merger that should immediately trigger a reconsideration of the merger and a reconsideration of the transfer of the broadcast licenses in a proceeding that is treated as a *de novo* review of the merger. Since Comcast has volunteered to give up its right to stop obeying a law in the event it is declared illegal or unconstitutional, it should have no problem giving up it right to challenge such a law.

FUNDAMENTAL REFORM IS LONG OVERDUE, FEDERAL AUTHORITIES SHOULD SEIZE THE MOMENT OF THE LARGEST MERGER IN HISTORY TO JUMP START THE REFORM PROCESS

Over the past quarter century there have been a few moments when a technology comes along that holds the possibility of breaking the chokehold that cable has on the multi-channel video programming market, but on each occasion policy mistakes were made that allowed the cable industry to strangle competition. This is the first big policy moment for determining whether the Internet will function as an alternative platform to compete with cable. If policymakers allow this merger to go forward without fundamental reform of the underlying industry structure, the prospects for a more competition-friendly, consumer-friendly multi-channel video marketplace will be dealt a severe setback.

It is only by taking the approach I have outlined that Federal authorities can do more than just preserve the current industry structure, which is riddled with anticompetitive and anti-consumer institutions and practices, that they can improve the terrain of the American video marketplace. This merger is an opportunity to jump-start the industry reform process.

THE ANTICOMPETITIVE EFFECTS OF THE COMCAST-NBC MERGER

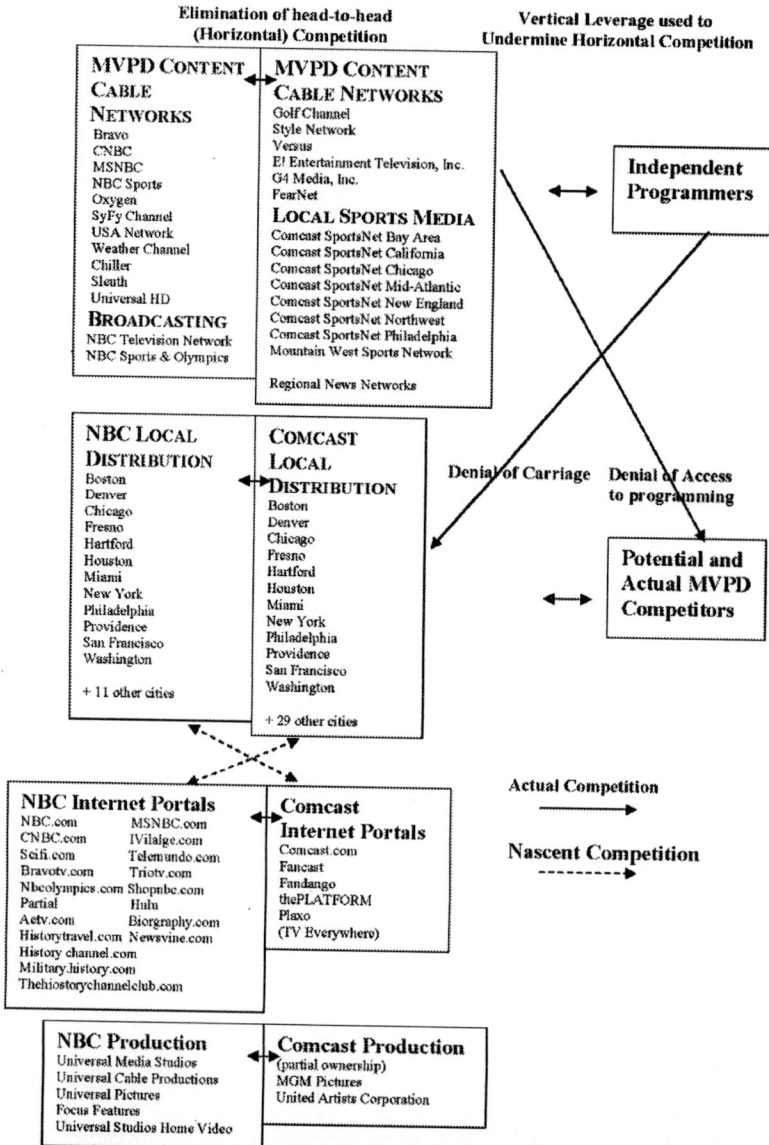

Elimination of head-to-head (Horizontal) Competition	Vertical Leverage used to Undermine Horizontal Competition

MVPD CONTENT CABLE NETWORKS
Bravo
CNBC
MSNBC
NBC Sports
Oxygen
SyFy Channel
USA Network
Weather Channel
Chiller
Sleuth
Universal HD
BROADCASTING
NBC Television Network
NBC Sports & Olympics

MVPD CONTENT CABLE NETWORKS
Golf Channel
Style Network
Versus
E! Entertainment Television, Inc.
G4 Media, Inc.
FearNet
LOCAL SPORTS MEDIA
Comcast SportsNet Bay Area
Comcast SportsNet California
Comcast SportsNet Chicago
Comcast SportsNet Mid-Atlantic
Comcast SportsNet New England
Comcast SportsNet Northwest
Comcast SportsNet Philadelphia
Mountain West Sports Network

Regional News Networks

Independent Programmers

NBC LOCAL DISTRIBUTION
Boston
Denver
Chicago
Fresno
Hartford
Houston
Miami
New York
Philadelphia
Providence
San Francisco
Washington

+ 11 other cities

COMCAST LOCAL DISTRIBUTION
Boston
Denver
Chicago
Fresno
Hartford
Houston
Miami
New York
Philadelphia
Providence
San Francisco
Washington

+ 29 other cities

Denial of Carriage   Denial of Access to programming

Potential and Actual MVPD Competitors

NBC Internet Portals
NBC.com   MSNBC.com
CNBC.com   IVilalge.com
Scifi.com   Telemundo.com
Bravotv.com   Triotv.com
Nbcolympics.com Shopnbc.com
Partial   Hulu
Aetv.com   Biorgraphy.com
Historytravel.com Newsvine.com
History channel.com
Military.history.com
Thehiostorychannelclub.com

Comcast Internet Portals
Comcast.com
Fancast
Fandango
thePLATFORM
Plaxo
(TV Everywhere)

Actual Competition
——————————→

Nascent Competition
- - - - - - - - - - →

NBC Production
Universal Media Studios
Universal Cable Productions
Universal Pictures
Focus Features
Universal Studios Home Video

Comcast Production
(partial ownership)
MGM Pictures
United Artists Corporation

I urge policymakers to think long and hard before they allow a merger that gives the parties incentives to harm competition and consumers, while increasing their ability to act on those incentives. This hearing should be the opening round in what must be a long and rigorous inquiry into a huge complex merger of immense importance to the American people. It should be the first step in a review process that concludes the merger is not in the public interest and should not be allowed to close.

ADDITIONAL LOCAL DISTRIBUTION

NBCU DISTRIBUTION

NBC Stations:
KNBC
Los Angeles
KNTV
San Jose/San Francisco
KXAS
Dallas/Fort Worth
WTVJ
San Diego
WNCN
Raleigh
WCMH
Columbus
WVTM
Birmingham
Telemundo Stations:
KVEA
Los Angeles
KWHY
Los Angeles
WSCV
Dallas/Fort Worth
KVDA
San Antonio
KDRX
Phoenix
KHRR
Tucson
WKAQ
Puerto Rico

COMCAST DISTRIBUTION

New Bedford
Springfield
Pittsburgh
Wilkes Barre
Baltimore
Richmond
Jacksonville
Orlando
West Palm Beach
Fort Myers
Tampa
Atlanta
Knoxville
Nashville
Chattanooga
Memphis
Peoria
Detroit
Grand Rapids
Indianapolis
Peoria
Champaign
Minneapolis/St. Paul
Albuquerque
Colorado Springs
Salt Lake City
Portland
Seattle
Sacramento

End Notes

[1] NBC Media Ownership Comments, FCC Docket 06-121 (filed Oct. 2006).

[2] Richard Epstein, "The Comcast and NBCU Merger: The Upside Down Analysis of Dr. Mark Cooper," *Perspectives from FSF Scholars,* 5:4, February 12, 2010.

[3] This critique of the Chicago School is amply documented in Robert Pitfosky (Ed.), *How the Chicago School Overshot the Mark: The Effects of Conservative Economic Analysis on U.S. Antitrust* (Oxford: Oxford University Press, 2008). On the under enforcement that results from the Chicago school approach see 6, 36, 244-247.

[4] *Id.,* at 5, 42, 57, 82.

[5] *Id.,* at 48, 52, 123.

[6] *Id.,* at 6,37-38, 85, 183.

[7] *Id.,* at 86, 127, 165.

[8] *Id.,* 79-81.

[9] *Id.,* at 5.

[10] *Id.,* at 42, 236.

[11] *Id.,* 243.

[12] *Id.,* 27, 57, 80, 126.

[13] *Id.,* at 5, 18, 42, 263.

[14] *Id.,* at 5.

[15] *Id.,* at 90, 263

[16] *Id.,* at 52, 127, 141.

[17] *Id.,* at 148-149.

[18] Mark Cooper, *Cable Mergers and Monopolies: Market Power in Digital Media and Communications Markets* (Washington, D.C.: Economic Policy Institute, 2002).

[19] *Id, at 93-97.*

[20] The charge that set off the implosion of the theory was ignited by Allan Greenspan's admission that there is a fundamental flaw in the theory. "Those of us who looked to the self-interest of lending institutions to protect shareholders' equity, myself included, are in a state of shocked disbelief. Such counterparty surveillance is a central pillar of our financial markets state of balance...If it fails, as occurred this year, market stability is undermined... "I made a mistake in presuming that the self-interests of organizations, specifically banks and others, were such that they were best capable of protecting their own shareholders and their equity in the firms (U. S. House of Representatives, Committee on Oversight and Government Reform, October 23, 2008) This has set off a series of analyses on all sides that retrospectively examine the cracks and weaknesses in the intellectual structure that should have been recognized (see for example Justin Fox, *The Myth of the Rational Market: A History ofRisk, Reward and Delusion on Wall Street* (New York: Harper Collins, 2009); Richard Posner, *A Failure of Capitalism: The Crisis of and the Decent Into Depression* (Cambridge: Harvard University Press, 2009); John Cassidy, *How Markets Fail the Logicfo Economic Calamities* New York: Farrar, Straus and Giroux, 2009). There were, of course, critics who recognized the problems much earlier, but whose warnings went unheeded (see for example, Joseph E. Stiglitz, *The Roaring Nineties* (New York: Norton, 2003); Robert Pollin, *Contours ofDescent: U.S. Economic Fractures and the Landscape ofAusterity* (Verso, 2005); Frank Portnoy, *Infectious Greed* (New York Holt, 2003); Robert Schiller, *Irrational Exuberance* (New York: Currency/Doublday, 2005);
Pitofsky, *op. cit.;* George Cooper, *The Origin ofFinancial Crises: Central Banks, Credit Bubbles and the Efficiency Market Fallacy* (New York: Vintage, 2008).

In: Media Industry Programming, Competition...   ISBN: 978-1-61122-078-0
Editors: Ryan E. Moore   ©2011 Nova Science Publishers, Inc.

Chapter 7

TESTIMONY OF LARRY COHEN, PRESIDENT, COMMUNICATIONS WORKERS OF AMERICA, BEFORE THE COMMITTEE TO THE JUDICIARY, HEARING ON "COMPETITION IN THE MEDIA AND ENTERTAINMENT DISTRIBUTION MARKET"

Larry Cohen

Good Morning, Mr. Chairman and Members of the Committee. Thank you for the opportunity to appear before you today. I am Larry Cohen, President of the Communications Workers of America. CWA represents more than 700,000 workers in communications, media, airlines, manufacturing, and the public sector. Specifically, we represent workers at both Comcast and NBC-Universal and can provide a unique perspective on the impact that this proposed merger would have on them and the industry.

The purpose of this hearing is to explore the potential anticompetitive impact of Comcast Corporation's proposed acquisition of NBC Universal. My testimony will focus on three areas: 1) the impact of the Comcast-NBC combination on jobs and the potential erosion of labor standards; 2) the anticompetitive behavior that currently pervades the video distribution and content markets and how that behavior will be exacerbated by this merger; and 3) the potential harms that such a transaction would pose to the emerging Internet video marketplace. At bottom, the public must be protected from the significant harms created by a combination of such unprecedented scale.

I. IMPACT ON WORKERS

The proposed Comcast acquisition of NBC poses considerable harm to workers. It likely will result in the loss of good jobs, the erosion of employee rights, and undermine living standards in the communications and media industries.

The new venture will be financially weaker the day after the merger. As part of the transaction, NBC debt will increase by approximately $8 billion. As a result, the new entity will be under intense pressure to cut costs and jobs. This is an all too familiar pattern in the media sector. Media companies over-leverage to pay for a merger, and then cut jobs to improve their balance sheets, only to discover that they do not have enough staff to produce quality news and entertainment programming. This in turn leads to a vicious cycle of declining audience share, less revenue, and even more cost-cutting. Absent firm commitments from Comcast and NBC to maintain current employment levels, there is no reason to believe that the Comcast/NBC joint venture will not follow this pattern. With official unemployment now at 10 percent, this is a time to evaluate all corporate transactions through a screen that assesses the impact on jobs. We should not support a corporate deal that would eliminate good jobs.

The communications and media sectors historically have been a source of good jobs for American workers, the result of more than 70 years of collective bargaining. But a Comcast acquisition of NBC would reverse this progress and undermine employment standards for workers in these sectors.

Comcast has adopted a low-road labor policy, one based on the violation of workers' rights. Comcast has a sordid track record of aggressive action to eliminate worker organization at companies that it has acquired.

In 2002, Comcast acquired AT&T Broadband. At the time, CWA represented about 5,000 cable employees there. After the transaction was announced, I met with Comcast executives and they made a commitment to me that they would respect their employees' right to a union voice on the job. Let me tell you what a Comcast commitment means. Soon after Comcast took control of AT&T Broadband, a senior vice president in Oregon announced: "We're going to wage war to decertify the CWA." And that is precisely what Comcast did in multiple locations.

Most of the organized units that Comcast acquired were in the process of negotiating a first contract. Comcast delayed bargaining for years, denied workers wage and benefit improvements provided to non-union employees, and supported decertification elections. Comcast refused to reach agreement on a first contract in 16 of the organized units that it acquired from AT&T.[1]

Comcast has fired and retaliated against workers that try to form a union. Before a union election, Comcast instructs its supervisors to ride along with technicians on the job, to meet repeatedly with workers one on one, and to hold mandatory meetings to convey its anti-union message.

CWA today represents Comcast employees in the Pittsburgh area. Comcast workers were forced to go through four union elections there in five years – three of them decertification attempts orchestrated by the company – before they finally won their union voice. Getting a first contract required overcoming further Comcast delaying tactics. Finally, Comcast has recognized that the workers there want a union voice and has negotiated a contract with CWA.

CWA represents Comcast employees in the San Francisco Bay and Detroit metropolitan areas. In both locations, Comcast has shifted about half the work to non-union lower-wage contractors, reducing secure jobs in areas hard-hit by unemployment.

Through these tactics, Comcast has managed to limit union representation to a small percentage of its workforce. Telecommunications has been a source of good jobs in this country, largely a result of more than 70 years of collective bargaining. The telecommunications industry has provided good jobs for women and minorities, with the result, as one economist wrote, that this industry is one of the few that has overcome market-based pay discrimination.[2] But Comcast – which competes directly with unionized telecom companies for voice, video, and broadband service – drags down the industry wage and benefit standards.

In contrast, NBC-Universal has a 70-year history of collective bargaining with multiple unions. To be sure, negotiations often deal with contentious issues, and the National Association of Broadcast Employees and Technicians (NABET) sector of CWA is currently in difficult negotiations with NBC-Universal on a contract covering technicians at the NBC network and stations in Washington, D.C., New York, Chicago, and Los Angeles. The contract expired 11 months ago. We are hopeful that we can resolve the issues that currently divide us. Although these are challenging negotiations, the bottom line is that NBC workers have a collective voice through their union – a right that Comcast has aggressively denied to their employees.

With the merger, an aggressively anti-union Comcast would be in control of labor relations, and an employer that has taken the low-road employment strategy will expand its ability to put downward pressure on living standards throughout the communications and media sectors.

Furthermore, the Comcast acquisition of NBC-Universal from General Electric represents a giant step backward on corporate governance practices. General Electric uses a one-share one-vote rule in shareholder voting. In contrast, Comcast has two classes of stock that gives super-majority voting rights to its CEO, Brian Roberts. Although Mr. Roberts owns only 1.23 percent of Comcast shares, he has 33 percent voting power. The Corporate Library, an independent shareholder research organization, has given Comcast an "F" on corporate governance practices. Comcast's undemocratic corporate governance structure mirrors its anti-democratic labor-management relations as well as its domination of the media marketplace.[3]

II. ANTICOMPETITIVE HARMS TO TODAY'S VIDEO MARKETPLACE

The proposed combination of Comcast, the nation's largest video service distributor, and NBC Universal, a leading video content producer, would create a vertically integrated entity with market power to increase cable rates, block competition in the video marketplace, and reduce jobs.

There is already too little competition in the video marketplace, as evidenced by the ever increasing rates that consumers pay year after year. The FCC estimates that from 1995 to 2008, the price of expanded basic service grew at three times the rate of inflation -- from $22.35 to $49.65, an increase of 122.1 percent, compared with an increase in the Consumer Price Index of 38.4 percent over the same period. (*See* chart, attached).[4] This merger would provide Comcast/NBC with added incentive and ability to engage in anti-competitive practices that would raise cable rates for consumers.

Today, competing video distributors are often forced to purchase large bundles of channels that they and their customers do not want. Following the merger, Comcast will have more premium content and will have the ability to bundle its less desirable cable channels with must-have NBC programming to secure higher rates and more favorable placement of its programming. This forced bundling will raise other video providers' costs, and those added costs translate into higher cable rates for consumers.

This is particularly problematic for small rural operators and new video competitors with a smaller subscriber base. Because Comcast and NBC give bulk discounts, they charge themselves less than they charge small and rural carriers on a per subscriber basis, raising the costs for cable subscription for customers of rural operators and new video entrants.

Today, some companies are trying to compete with incumbent cable operators, investing significant resources to build out their networks and enter the video marketplace. This merger would provide Comcast/NBC with the incentive and ability to block or limit that competition, and block or limit the investment and jobs that accompany those efforts.

As competitors' costs increase, those companies trying to compete will invest less in building out their networks and hire fewer people. As a result of this proposed merger, Comcast/NBCU will have the market power to stifle competitive entry by new video operators. As a result, there will be fewer companies competing to provide traditional cable video services, fewer choices and higher prices for consumers, and lost jobs from these potential competitors.

In the past, Comcast has used its ownership of sports programming in an anti-competitive way. For example, Comcast has prevented DirecTV and Dish Network from accessing its SportsNet Philadelphia channel, which carries the games of Major League Baseball's Phillies, the NBA's Sixers and the NHL's Flyers. (Comcast has a controlling interest in the Sixers and Flyers.) By withholding the games of the three Philadelphia professional sports teams from its rivals, Comcast has had a powerful marketing advantage against satellite TV competitors.

Comcast has faced numerous FCC complaints from programmers for discrimination and anti-competitive behavior. The NFL Network, the Tennis Channel, MASN, a regional sports network, and Wealth TV, an emerging HD programmer, have filed formal FCC complaints against Comcast. These complaints allege that Comcast carried its own programming on favorable terms while refusing to carry independent programming on equal terms – or to carry such programming at all. Should regulators approve the Comcast-NBCU merger, Comcast will have more affiliated content and even more of an incentive to favor its own programming in its carriage decisions.

This may result in Comcast refusing to carry competitors' programming, paying them less for carriage, or placing them on a program tier with fewer viewers. After acquiring NBC programming, Comcast will have even greater incentives to favor its own array of programming, shutting out the independent voices of other programmers, leaving consumers with less quality, choice and diversity in programming. In fact, Comcast Cable's President and COO Steve Burke made remarks during the NFL's program carriage complaint with the FCC that Comcast treats affiliate networks "like siblings as opposed to strangers."

Meanwhile, bringing a carriage access complaint to the FCC is not a meaningful remedy. The complaint process currently lacks any concrete deadlines for FCC action, with many complaints languishing at the Commission for years.

Today, Comcast's regional and local programming networks compete with NBC's owned and operated stations for news and entertainment programming. A merged Comcast/NBCU would have the incentive to merge these operations, reducing quality, diversity, competition, and employment in video programming. Already, NBC has pioneered local news sharing agreements that in effect merge NBC's local news gathering with those of its broadcast competitors. Under these arrangements, NBC and its former competitors jointly determine news assignments and crew assignments, replacing what were once competing news operations with shared news gathering. In New York City, for example, six stations owned by four different owners (including NBC's owned and operated station and its Telemundo station) cooperate in a local news sharing venture.

This merger also threatens to eliminate a current competitor for local advertising. Contrary to Comcast/NBC claims today, Comcast is a major and sometimes even the most significant competitor for local ad dollars in some local advertising markets. In 2006, NBC made this very argument to the Federal Communications Commission. NBC stated that cable's local advertising dollars exceeded the total advertising dollars at NBC local stations, and actually were greater than advertising revenues at the number one ranked station in several markets, including Philadelphia (greater than the ABC station by $26 million) and San Francisco (greater than the Fox station by $70 million).[5] Comcast's local ad share has grown since then. In addition, an independent broadcaster will not be able to offer the volume discounts and package deals for advertising across dozens of channels that the merged entity will be able to do.

This translates into less revenue for competing broadcasters to produce local news and to hire workers. As a result, broadcasters will no doubt be forced to scale back local news production, with negative impact on diversity, competition, and adequate staffing that drive quality news.

In terms of local market share, Comcast's ad penetration is analogous to NBC trying to merge with the number one ranked station in the market, a practice prohibited by the FCC under its dual network rule. That rule permits common ownership of multiple broadcast networks but prohibits a merger of the "top four" networks, i.e., ABC, CBS, Fox, and NBC.

III. ANTICOMPETITIVE HARMS IN EMERGING ONLINE VIDEO MARKET

Another area of concern posed by the Comcast-NBC Universal merger is in the developing online video marketplace. New entrants are beginning to offer a number of video streaming services on the Internet and "over the top" services that bring Internet video directly to the television. This premium content that is available online increases the value of broadband subscription to consumers. Thus, the availability and ease of accessing video online is an important means to encourage the deployment and adoption of broadband. And as broadband adoption increases, some users are able to choose to "cut the cable cord," canceling their cable subscription and relying on the Internet for television. In fact, the FCC recently concluded that internet video and video devices are an important part of developing a National Broadband Plan.[6]

The Comcast-NBC merger has the potential to bring this to a halt by limiting the ability of over-the-top service providers to provide video. A combined Comcast/NBC could limit consumers' online access to NBC content altogether or charge consumers higher prices to access that content unless they already subscribe to cable services. This is the TV Everywhere model that Comcast and NBC have already begun to deploy, bundling content with cable subscription, thereby forcing internet customers to buy cable packages in order to see content online from NBC.

TV Everywhere is an initiative being pursued by a number of cable companies, but led by Comcast. Under the TV Everywhere model, Comcast video subscribers have access to video content online for free, just as they do today. Online consumers, however, are forced to pay higher rates or restricted from accessing the content at all. For example, that is what is happening with some Olympic coverage from NBC today. In the biggest TV Everywhere trial,

NBC restricts access to live streaming and full replay of Olympic events to consumers who can "authenticate" that they are paying cable subscribers.

TV Everywhere creates a mechanism for programmers and content providers to have a "walled garden" of online video content, only available to those who pay their monthly cable subscriptions. In doing so, TV Everywhere denies independent video distributors access to must-have programming, and creates a barrier to entry in the video distribution market for Internet-only video distributors. This extension and protection of the cable business model effectively "cablizes" the Internet as we know it today, thus diminishing innovation, depressing investment in broadband and ultimately eliminating jobs.

In the end, consumers lose out. TV Everywhere would protect the cable business model by imposing its subscription pricing structure on the Internet. Where customers have traditionally accessed content for free, they would now be forced to pay. Where the internet use to be a source of expanding consumer choice and diversity of programming content, it would be used to protect the current cable incumbents.

A merged Comcast/NBC would have the ability to force this business model on other distributors through their ownership of NBC's content. Today, NBC owns a 30 percent interest in a website called Hulu.com that offers free, advertising-supported streaming video of broadcast and cable television shows and movies. In acquiring NBC, Comcast would secure a substantial interest in Hulu.com, which is the second leading online video provider. As a result, Comcast would play a critical role in the public's ability to continue accessing the Internet's growing video services.

All of the actions I have just described restrict the Internet from developing into a platform for competitive video alternatives. These actions in essence protect the cable-channel business platform at the expense of new video entrants, thereby devaluing the broadband investment of competitive companies. The end result is that companies will invest less in broadband deployment, put less fiber in the ground and hire fewer people.

CONCLUSION

The Comcast/NBC merger's potential to limit growth, investment and jobs is not in the public interest. Given its anticompetitive and anti-consumer effects, federal regulators cannot pass this merger without carefully considering the significant impact the merging companies will have on video competition, choice and jobs. Moreover, federal regulators cannot rely on the voluntary public interest commitments offered by Comcast and NBC Universal alone. The voluntary commitments are: 1) insufficient to address adequately the very real competitive harms; and 2) in many cases, rest on pending actions before federal regulators. As a result, prior to addressing this merger, CWA believes that both the DOJ and the FCC should complete many of the actions that will address some of these issues from a broader industry-wide perspective. Federal regulators would then have the ability to craft any additional, specific merger conditions that are necessary to further address the potential harms caused by this combination.

Again, I want to thank the Committee for giving me the opportunity to testify today and for Chairman Conyer's leadership on this issue. I look forward to answering any questions that the Members of the Committee may have.

Attachement

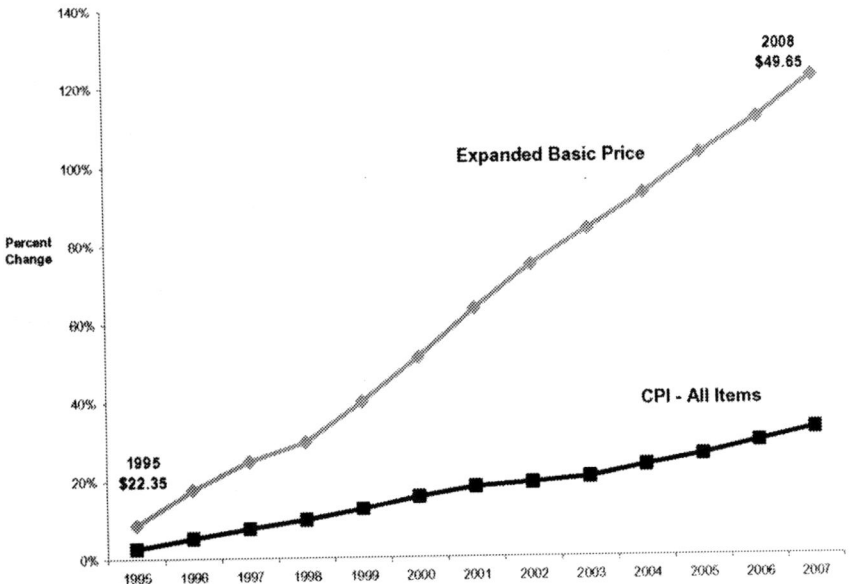

Source: FCC, Report on Cable Prices, MM Docket No. 92-266, Chart 1, 2009

Cable Prices Increase at Three Times the Rate of Inflation Cable Price and the CPI, 1995-2008

Citations of Online Reports

[1] American Rights at Work, *No Bargain: Comcast and the Future of Workers' Rights in Telecommunications,*" 2004 (available at http://www.americanrightsatwork.org/publications/general/no-bargain-comcast-and-the-futureof- workers-rights-in-telecommunication.html)

[2] Vicky Lovell, Heidi Hartmann, Jessica Koski. (2006). *Making the Right Call: Jobs and Diversity in the Communications and Media Sector,* Washington, D.C.: Institute for Women's Policy Research, 2006 (available at http://www.iwpr.org/pdf/C364.pdf)

End Notes

[1] American Rights at Work, *No Bargain: Comcast and the Future of Workers' Rights in Telecommunications,*" 2004 (available at *http://www.americanrightsatwork.org/* publications/general/no-bargain-comcast-and-the-future-of-workers-rights-in-telecommunication.html).

[2] Vicky Lovell, Heidi Hartmann, Jessica Koski, *Making the Right Call: Jobs and Diversity in the Communications and Media Sector,* Washington, D.C.: Institute for Women's Policy Research, 2006 (available at http://www.iwpr.org/pdf/C364.pdf).

[3] The Corporate Library, Comcast Corporate Governance Report, Feb. 23, 2010.

[4] *In the Matter of Implementation of Section 3 of the Cable Television Consumer Protection and Competition Act of 1992,* Report on Cable Industry Prices, MM Docket No. 92-266 ¶ 2, Chart 1 (2009).

[5] NBC Media Ownership Comments, FCC 06-121 (Oct. 2006).

[6] *Comment Sought on Video Device Innovation,* NBP Public Notice #27, GN Docket Nos. 09-47, 09-51, 09-137; CS Docket No. 97-80 (Dec. 21, 2009).

In: Media Industry Programming, Competition…   ISBN: 978-1-61122-078-0
Editors: Ryan E. Moore   ©2011 Nova Science Publishers, Inc.

Chapter 8

TESTIMONY OF ANDREW JAY SCHWARTZMAN, PRESIDENT AND CEO, MEDIA ACCESS PROJECT, BEFORE THE COMMITTEE TO THE JUDICIARY, HEARING ON "COMPETITION IN THE MEDIA AND ENTERTAINMENT DISTRIBUTION MARKET"

Andrew Jay Schwartzman

I believe that Comcast should not be allowed to acquire NBC Universal.

As I said when the Comcast/NBCU transaction was announced, this is the most important media merger since Lucy met Desi. Comcast seeks to combine its huge cable and internet footprint, reaching about 30 % of the nation's homes, with NBCU's gigantic content assets. NBCU has 26 TV stations in the country's largest markets, the NBC network, several of the highest rated cable TV networks and the Universal film library.

WHY THIS IS SUCH AN IMPORTANT TRANSACTION

At the outset, I want to stress that my opposition to the Comcast/NBCU merger is not based on animus. Brian Roberts is not evil; to the contrary, he is a public spirited, ethical businessman. Even though I have problems with his labor/management practices and his corporate governance structure, I recognize that he is motivated by business considerations and not some sort of design to undermine American democracy.

But the consequences of this deal nonetheless could have precisely that effect.

Concentration of control in the mass media poses unique questions for policymakers and regulators. Unlike any other line of business, media properties raise important questions which go to the very nature of democratic self-governance. Our viewpoints and perspectives on political and social issues are the outgrowth of what we hear and watch. Indeed, it has been clear for some 60 years that antitrust principles overlap with First Amendment doctrine. The seminal case in this regard is *United States v. Associated Press,* where the Supreme Court applied the Sherman Act to newspapers.

Writing for the majority in *Associated Press,* Justice Black held that the First Amendment provided powerful support for applying the Sherman Act because it "rests on the assumption that the widest possible dissemination of information from diverse and antagonistic sources is essential to the welfare of the public...." Justice Frankfurter emphasized in his concurring opinion that the case was about a commodity more important than peanuts or potatoes, that it was about who we are as a nation. "A free press," he said, "is indispensable to the workings of our democratic society." For that reason, he wrote, "the incidence of restraints upon the promotion of truth through denial of access to the basis for understanding calls into play considerations very different form comparable restraints in a cooperative enterprise having merely a commercial aspect."

A notable example of how this concept has been applied in practice can be found in Judge Greene's treatment of the AT&T consent decree. In imposing restrictions on what was then described as "electronic publishing," he held that both competitive and First Amendment considerations separately supported his action.

Judge Greene made clear that application of these objectives is not delimited to Title III of the Communications Act. "Certainly," he said, "the Court does not here sit to decide on the allocation of broadcast licenses. Yet, like the FCC, it is called upon to make a judgment with respect to the public interest and, like the FCC, it must make that decision with respect to a regulated industry and a regulated company." Thus, he said, it was necessary for him to "take into account the decree's effect on other public policies, such as the First Amendment principle of diversity in dissemination of information to the American public. Consideration of this policy is especially appropriate because, as the Supreme Court has recognized, with respect to promoting diversity in sources of information, the values underlying the First Amendment coincide with the policy of the antitrust laws." *Id.*

Time precludes extensive discussion, so today I will emphasize just three of the many anti-competitive ways in which Comcast could leverage ownership of NBC content assets to extend its reach in distribution of video programming and Internet services. My focus on national issues does not mean that I am unconcerned about the impact of Comcast's plans on the communities where it will own both TV stations and cable systems. Rather, it means that I know that my friend Mark Cooper is going to address this question extensively in his remarks, with which I wish to associate myself.

A COMCAST/NBCU COMBINATION WILL HARM INDEPENDENT PROGRAMMERS AND THE PUBLIC

First, I want to address how approval of this merger would increase Comcast's power to squeeze out independent programmers with diverse editorial perspectives.

There are scores of cable networks which have been unable to obtain carriage on Comcast and other cable systems. I'm here, and they are not, because some of these companies have told me that they are afraid of retaliation. Indeed, over the last several years numerous programmers such as NFL Network and WealthTV have unsuccessfully pursued carriage complaints at the FCC. In each case, they argued that Comcast favored its own channels while refusing to carry independent programming on workable terms, if at all. Acquisition of NBC's stable of cable networks will greatly exacerbate the imbalance of power.

If Comcast is permitted to purchase the NBC TV stations and its highly viewed cable networks, Comcast will be able to bundle its programming when it seeks carriage deals with other multichannel video programming distributors ("MVPDs") such as telephone and satellite companies. This enables Comcast to obtain distribution for new and secondary channels which otherwise would never receive such treatment. Each time a Comcast channel is forced into the program menu, there is one less slot for independently owned programming.

The problem is even greater with respect to carriage on Comcast's own cable systems. The existing legal framework already gives Comcast every incentive to favor its own programming over independently produced cable channels. This can include refusal to carry competitors, paying them far less for carriage or placing them on a lesser watched program tier.

After the acquisition, Comcast will have even more cable networks to favor in deciding what to carry on its cable platform. Because it will create incentives for Comcast to make programming decisions based on self-serving financial factors rather than program quality, approval of the merger would mean that the public will get inferior programming. Discrimination of this kind also generates higher prices for all Americans, not just Comcast customers. Since Comcast will be paying itself for program carriage, it can set a higher wholesale price for its programming, so that competing MVPDs will also have to pay higher prices. This, of course, will be passed on to their customers.

There ought to be a law against such abuse. In fact, there is. Section 616 of the Communications Act prohibits cable companies from demanding an equity interest in a programmer or exclusivity rights as a condition for carriage. It also prohibits cable companies from discriminating in favor of their own programming.

Comcast understandably, but unpersuasively, argues that existing laws and regulations sufficiently protect independent programmers and the public. Once those of us in the public interest community called attention to the fact that Comcast has argued that enforcement of Section 616 is unconstitutional, Comcast has suggested, but not quite promised, that it will not pursue such a challenge in the future. However, this does not change a more fundamental fact, which is that the existing statute does not work. The cost of litigating program carriage cases has proven to be prohibitive, and the FCC has adopted almost insuperable legal hurdles for complainants to overcome. Since Section 616 was enacted in 1992, only a handful of complaints have made it past the threshold level. There is no time limit for FCC action, and complaints and appeals often have been stalled at the FCC for months and years. Even when there is FCC action, the reward for success is a lengthy and expensive legal trial with the legal deck stacked in favor of the cable companies.

A case in point is the difference in treatment between the MLB Network and the NFL Network. For more than a decade, the National Football League's NFL Network has fought for carriage on widely viewed cable tiers at fair prices. It has been unable to reach agreements with a number of major cable operators. By contrast, Versus, a competing but far less viewed sports channel owned by Comcast, has been placed on a basic tier. Finally, the NFL filed a Section 616 complaint against Comcast, alleging that Comcast would not place the NFL Network on the same tier that Comcast placed its own sports networks and that it had conditioned its willingness to carry the NFL Network upon receipt of a financial interest in NFL programming. After considerable delay, the FCC finally directed that a hearing be held. Eventually, a year after its complaint was filed, the delay and cost of the hearing forced the NFL to accept a settlement which provided inferior channel placement at a price far below what the NFL had sought. Even the NFL, with its vast resources, couldn't crack the Comcast stranglehold without lawsuits, FCC proceedings, and years of uncertainty before it reached a negotiated settlement which was less than what it wanted.

Major League Baseball learned from the NFL's experience, and took a different tack. When it created the MLB Network it did what the NFL has refused to do, and offered significant ownership interests to the major cable operators, including Comcast. Not surprisingly, from the moment of its launch, the MLB Network has been carried on the basic cable tier.

Plainly, existing law does not provide adequate protection for independent programmers. Acquisition of the NBCU program networks will only make things worse.

THE PROPOSED MERGER WILL HURT OTHER DISTRIBUTORS AND THE PUBLIC

Combining NBC and Universal content with Comcast's cable and Internet distribution systems will also give the merged company vastly increased power over content distribution markets. Depending on the circumstance, Comcast could choose to withhold its programming or force it upon competitors at inflated prices. This in turn will increase cable bills and deprive customers of access to programming from diverse sources.

There are countless ways in which Comcast could exercise such leverage. For example, it can condition the sale of important "must have" programming, including that of the NBC and Tele-mundo TV stations, upon acceptance of undesired, secondary channels which would never receive carriage in a competitive market. Or it could withhold or delay access to the Universal film library from competing MVPDs, or demand vastly inflated licensing fees.

As with the program carriage problem discussed above, Comcast would assure you that existing law is sufficient to protect against harm. Indeed, the "program access" provision in Section 628 of the Communications Act requires vertically integrated cable operators to share their programming with competitors without discrimination in prices, terms or conditions of sale. Moreover, the Commission has recently closed, in part, the so-called "terrestrial loophole" that has permitted Comcast and other cable companies withhold regional programming, such as the Comcast Sports Network in Philadelphia.

Comcast has also brought a legal challenge the FCC's legal authority to continue enforcing program access rules' ban on exclusive contracts. Although Mr. Roberts has more recently said that Comcast is willing to consider a promise to adhere to the rules regardless of the outcome of its court case, he has thus far refused to drop the lawsuit.

That aside, there are many reasons why existing law is insufficient to protect Comcast's competitors and their customers. First, even if Comcast doesn't upset them in court, the program access rules expire in two years, and there is no assurance that they will be extended. In any event, the program access regime does not preclude bundling, which is one of the principal anti-competitive mechanisms Comcast is likely to employ. Although Section 628 prohibits discrimination against competitors, this simply means that as long as Comcast overcharges itself, it can overcharge everyone else. In addition, the program access provision does not apply to a large proportion of the content that Comcast is acquiring, such as feature films and other video on demand content. Moreover, Section 628 is a right without a remedy; the FCC's complaint process is so onerous and time consuming that I am unaware of a single program access complaint which has ever been granted. And, no less importantly, the negotiation process is one-sided. There is no "standstill" requirement, so that when a carriage agreement expires, all of the power is in the hands of the programmer.

Retransmission consent for NBC Network and Telemundo programming poses another especially important problem. Without Comcast's permission, competing MVPDs would be unable to offer this essential programming. As the recent Fox/Time Warner Cable dispute demonstrated, even the most powerful satellite or cable companies cannot last for a day without major TV network programming. Post-merger, Comcast could decide to pay itself twice the fair value for NBC and Telemundo programming and then turn around and exact the same inflated price from its competitors, who would be forced to pass on the overcharges to their customers. Or, Comcast could tie the carriage of this programming to the carriage, at favorable channel locations, of the least desirable of its cable channels, also at inflated prices.

Existing retransmission consent rules are an unreliable tool for assisting Comcast's video competitors. Section 325 of the Communications Act supposedly mandates "good faith" negotiation, but it does not prohibit price or packaging discrimination; it simply requires a commercially feasible offer. NBC already requires MVPDs to accept bundles of cable programming in order to get the NBC and Telemundo programming; addition of the Comcast distribution magnifies the leverage by several orders of magnitude. The FCC's complaint process offers no effective remedies other than a finding that one party has acted badly. There is no time limit for FCC action, and as with program access, there is no "standstill" provision to maintain some level of parity during negotiations.

In this connection, I would observe that Mr. Roberts has recently indicated that Comcast may be willing to increase retransmission payments to NBC affiliates. This may or may not be a good thing for the future of broadcast TV, but no one should doubt that the impact of this would be to raise cable rates for everyone. If Comcast pays more for retransmission consent, customary industry contractual arrangements are such that the same higher rates will be applied to affiliates of other networks as well. Mr. Roberts certainly is not going to absorb those costs; they will be passed on to the public in the form of higher rates.

HOW COULD THIS HAPPEN?

One could fairly ask how Congress could have created a system which would permit a single company to operate cable systems, cable TV networks and a stable of owned-and operated TV stations, much less a major TV network? The answer is that Congress never contemplated such a combination. When the program access and retransmission consent provisions were enacted in 1992, the law already prohibited common ownership of a cable system and a local TV station. The local cable/television cross-ownership rule was eliminated a few years ago by judicial action, not legislation. There is very little doubt that Congress would have included much stronger protections if it ever thought that such cross-ownership would ever be permitted.

COMCAST'S ACQUISITION OF NBCU WOULD JEOPARDIZE DEVELOPMENT OF A FREE AND OPEN INTERNET

Internet technology offers the prospect of creating vibrant and highly competitive markets for video programming. Members of the public can, or will soon be able to, receive high-definition video via the Internet. Comcast has already taken steps to kill off such competition, and acquisition of NBC's content will greatly enhance that campaign.

Comcast has strong reason to keep its customers from migrating to existing and potential Internet-delivered video competitors. Control of NBCU branded content as well as its one-third interest in Hulu would give Comcast a powerful mechanism to retain its video services revenue stream by strangling potential Internet-based competition before it can even get off the ground.

It would be reasonable to expect that the public's reaction to the diminished choice and increased prices resulting from a Comcast/NBCU merger would be to seek alternative ways to obtain video content. The possibility that viewers may soon be able to watch Internet-based video on a TV becomes by simply clicking on a remote control, Comcast's business model will be threatened. Indeed, a brave few have already decided to "cut the cord" by cancelling their cable TV service and relying on the increasing amount of content available over the air and on the Internet. It is becoming ever easier to connect digital TV sets directly to the Internet and employ services like iTunes, and Hulu and devices like Boxee and Roku, while relying on free over-the-air television for news and other local programming. Existing online-only video distributors such as Vudu and Netflix are growing rapidly. There is no technological or business model reason why there won't soon be Internet-delivered "virtual cable" services with a menu offering the popular "linear programming," including the major cable TV networks. Indeed, the *Wall Street Journal* recently reported that a company called Move Networks has discussed plans to offer just such a service. (I would be a little more optimistic about the prospects for Move Networks' becoming a competitive offering if Comcast were not a major customer of, and investor in, that company.)

This is an existential threat to the cable industry. Its answer is "TV Everywhere." Comcast's version, which goes under the unwieldy name of "Fancast XFinity," offers the superficially attractive opportunity for its video and Internet customers to view video over the Internet without extra charge.

The catch, which is a very big catch indeed, is that you must keep your cable TV subscription.

XFinity represents an attempt to kill off potential competition while preserving the cable TV revenue stream indefinitely. XFinity is available only in Comcast regions, as it and other cable operators have continued their longstanding tacit agreement of never competing with each other on price or services. And, while we are told that satellite and telco competitors will soon be allowed to offer Comcast's content, the same opportunity will not be offered to any new online-only TV distributors. Nor is it clear that this content will be made available under the same terms and conditions. By design, XFinity cuts off the flow of programming to disruptive new entrants.

The XFinity offering also threatens existing independent programmers. Comcast has conditioned cable TV carriage on contractual provisions which prevent programmers from selling their content to competing online distributors at least temporarily and, perhaps, permanently.

Last month, MAP joined with Free Press and other public interest groups in issuing a white paper which set forth in detail how the cable industry has colluded to create the "TV Everywhere" model. As the report says,

> Stripped of slick marketing, TV Everywhere consists of agreements among competitors to divide markets, raise prices, exclude new competitors and tie products.

Comcast's acquisition of NBCU's programming vastly increases its leverage to force XFinity upon its customers and to stifle new competitors. All of the program carriage and program access problems that video competitors currently face will be replicated in the Internet space, but there are no similar legal protections. Of particular note in this regard is the fact that NBC has a major ownership interest in Hulu, the leading Internet video service. If it is in Comcast's interest, it can cripple Hulu by withdrawing NBC content from Hulu. Alternatively, Comcast may choose to make the NBC content exclusive to Hulu and withhold it from new Internet-delivered video competitors.

Comcast's control of the vast Universal film library would be another important building block in the effort to stifle new Internet competitors. Comcast can withhold these products from Internet competitors or delay their availability while offering them exclusively on XFinity. For example, it could target DishTV, which competes in the video market. Dish has an Internet delivered video service called Dish Online. By denying Universal's film library to Dish Online, Comcast could drive video customers to itself. If Dish were uncooperative, Comcast could also deny XFinity to Dish in Comcast markets.

Finally, while I hope that the FCC quickly moves to adopt "Network Neutrality" rules to prohibit discrimination in delivery of broadband services, I must point out that, in the absence of such provisions, Comcast can degrade or otherwise discriminate against competitors seeking to deliver Internet video program services to Comcast's Internet customers.

CONCLUSION

Comcast's proposed acquisition of NBCU is profoundly anti-competitive and will adversely affect the marketplace of ideas as well. I hope the Committee members will join Media Access Project in urging the FCC and the Department of Justice to block it.

CHAPTER SOURCES

The following chapters have been previously published:

Chapter 1 – This is an edited reformatted and augmented version of a United States Government Accountability Office publication, report GAO-10-369, dated March 2010.

Chapter 2 – This is an edited reformatted and augmented version of a Congressional Research Service publication, report R41274, dated June 4, 2010.

Chapter 3 – This is an edited reformatted and augmented version of a Congressional Research Service publication, report RL32460, dated February 1, 2010.

Chapter 4 – These remarks were delivered as testimony given on February 25, 2010. Brian L. Roberts, Chairman and Chief Executive Officer, Comcast Corporation and Jeff Zucker, President and Chief Executive Officer, NBC Universal, before the Committee on the Judiciary, United States House of Representitatives.

Chapter 5 – These remarks were delivered as testimony given on February 25, 2010. Thomas W. Hazlett, Panel on the Comcast-NBCU Venture, before the United States House of Representitatives, Judiciary Committtee Hearings.

Chapter 6 - These remarks were delivered as testimony given on February 25, 2010. Mark Cooper, Director of Research, Consumer Federation of Amercica on behalf of Consumer Federation of America free Press Consumers Union, before the United States House of Representatives, Committee on the Judiciary, Subcommittee on Antitrust, Competition Policy and Consumer Rights.

Chapter 7 - These remarks were delivered as testimony given on February 25, 2010. Larry Cohen, President, Communications Workers of America, before the United States House of Representatives, Committee on the Judiciary.

Chapter 8 - These remarks were delivered as testimony given on February 25, 2010. Andrew Jay Schwartzman, president and CEO, Media Access Project, before the Committee on the Judciary, United States House of Represenatives.

INDEX